Embodiment, Aesthetics, and Ethics in Romantic Literature and Science

Romantic Reconfigurations:
Studies in Literature and Culture 1780–1850

Series Editors:
Professor Tim Fulford, De Montfort University
Professor Alan Vardy, Hunter College and the Graduate Center, CUNY

As befits a series published in the city of Roscoe and Rushton, a city that linked Britain to the transatlantic trade in cotton, in sugar, and in people, *Romantic Reconfigurations* reconfigures the literary and cultural geographies and histories of Romanticism. Topics featured include, but are by no means confined to, provincial and labouring-class writing, diasporic and colonial writing, natural history and other scientific discourse, journalism, popular culture, music and theatre, landscape and nature, cosmopolitanism and travel, poetics and form.

Embodiment, Aesthetics, and Ethics in Romantic Literature and Science

Lisa Ann Robertson

LIVERPOOL UNIVERSITY PRESS

First published 2025 by
Liverpool University Press
4 Cambridge Street
Liverpool
L69 7ZU

British Library Cataloguing-in-Publication data
A British Library CIP record is available

ISBN: 978-1-83624-485-1 (hardback)
eISBN: 978-1-83624-872-9 (pdf)
eISBN: 978-1-83624-890-3 (epub)

Typeset by Carnegie Book Production, Lancaster
Printed and bound by CPI Group (UK) Ltd, Croydon CR0 4YY

Contents

My first book is dedicated to Marlene Longenecker and David Miall, who are now making their diurnal rounds.

Acknowledgements

The acknowledgements section provides an opportunity to reflect on the unseen labour that often falls to women and without which no academic monograph could come to fruition. I am, of course, talking about the domestic chores that are time consuming, tedious, and absolutely crucial for the smooth functioning of every aspect of life, especially writing a book. Thus, I am particularly grateful for the support I've received from my life partner, Martin Prendergast, to whom I owe my deepest debt of gratitude for assisting me in this most material of ways. At crucial phases of writing and revising (of which there were many), food appeared on my desk, dirty plates were whisked away, and I never wanted for coffee despite not setting foot inside a grocery store for weeks on end. The moral support Martin provided me was also invaluable. His confidence in me never wavered, and he knew that I would finish this book even when I was not so sure. Thank you, my love, for creating the material conditions that enabled this book to come into being.

Because this book was so many, many years in the making, the people to whom I owe thanks are legion. To trace it back to its earliest beginnings, this project would not have gained traction without the guidance of my doctoral supervisor, David Miall, my dissertation committee James Mulvihill and Stan Ruecker, and Alan Richardson, my external reader. They were instrumental during my University of Alberta days in helping me to conceptualize, research, and write up the first glimmerings of a book that is most decidedly *not* my dissertation.

In 2013, Adam Komisaruk organized a seminar on the science of the Romantic imagination, where I met Richard C. Sha, who led the seminar and to whom I owe an immense intellectual debt. Known for his generosity, Richard had countless conversations with me, commented on conference presentations, and read chapter drafts. He patiently waited for me to figure out what I was arguing, which I suspect he knew long before I did. Richard never hesitated to ask the sharpest, most useful questions, such as, 'So what? Why should I care?!' I am grateful for this constructive and honest criticism. Thank you, Richard, for your investment of time and energy, and for making this book much better than it would have been without your mentorship. Any flaws in the finished product are surely instances where I did not take his advice.

The manuscript version of this book has had many readers from its earliest form to the pages that you have before you. In my graduate student days, I benefitted from the feedback and friendship of Louise Harrington, Clare Mulcahy, and Brianna Wells. Thanks to them, I no longer torment people with two-page paragraphs, and I explain contemporary cognitive concepts more clearly. At the University of South Dakota (USD), several people read an early and entirely too long version of my Coleridge chapter, including Prentiss Clark, John Dudley, Darlene Farabee, Paul Formisano, Benjamin Hagen, Sara Lampert, and Skip Willman. Skip's suggestion to consult Niklas Luhmann paid off, despite my initial resistance. I also benefitted from Jeffrey W. Barbeau's comments on an article for *The Coleridge Bulletin*, portions of which appear in this book. Sara and Ben provided ongoing feedback as I wrangled it, and other chapters, into something manageable.

As part of a seminar sponsored by the North American Society for the Study of Romanticism's (NASSR) Science, Medicine, and Psychology Caucus organized by Dahlia Porter, Joel Faflak responded to an early draft of the Darwin chapter. I am indebted to him for helping me recognize the theoretical potential of posthumanism for framing Darwin's organicism. Dahlia's input sharpened the chapter as I revised a portion of it for publication in *European Romantic Review* and as I developed it afterwards. Thank you, Dahlia, for being a skilled editor of my work.

Many others have been instrumental in developing my thinking along the way. Sharon Ruston's and Tim Fulford's expertise on science and Romanticism, and the workshops and conferences they have organized and led, shaped the way I think about organicism. Mary Fairclough's feedback on an earlier manuscript made this one much better, as did the input of many others with whom I've presented and chatted over the years. A host of formal and informal networks that enable academics to gather, share, and discuss their work has directly and indirectly improved the calibre of my thinking. I owe thanks to the USD graduate students in my seminars on 'Mind and Body in Romanticism' and to the students and faculty who attended my colloquium talks, which gave me a forum for presenting my work in progress. Several chapters were developed in this way. Christabel Scaife made the submission and editorial process smooth, while suggestions from my two anonymous reviewers improved the historical context and scholarly framing of this monograph.

Bridget Diamond-Welch, Sara Lampert, Prentiss Clark, Shannon Cole, and Heather Love provided fits of laughter, pointed political and feminist critique (some might call it kvetching), and delicious gatherings involving food, friendship, and creative activity that provided just enough life to balance out the work. They helped energize me throughout the process. Finally, I would be remiss if I did not acknowledge my cats—Camilo, who ensured that my arms did not atrophy by lying on them while I typed, and Lilabet, who made sure that I got at least minimal sleep by herding me to bed when she determined I had been up working late enough. Without all the beings with whom I am entangled in an interconnected web, this book would not be. Many thanks and much love to you all.

Introduction

Revisiting Romantic Organicism

Seamus Perry aptly describes the critical attitude towards Romantic organicism when he asserts that 'Organicism has regularly served as a romantic bogey in recent criticism'.[1] Over the past forty years, in the wake of deconstruction and New Historicism, it has not fared well as a literary concept. Yet scholars who embraced organicism, such as the early twentieth-century New Critics and mid-century Romanticists, defined it in ways that not only justified later critiques but also deprived it of its critical due. Recuperating Romantic organicism is the goal of this monograph, which I accomplish by resituating it in its historical context while bringing to bear what Marjorie Levinson calls '"post-classical" frameworks in the physical and life sciences' that offer new analytical tools that were unavailable until the beginning of the twenty-first century.[2] Organicism's shift from apotheosis to scapegoat over the course of the twentieth century has a clear trajectory. The attack on organicism began as a critique of the New Critics' reification of poetry as a 'form of knowledge that belongs to a different order of being' from the merely empirical.[3] In the 1930s and 40s, scholars such as Cleanth Brooks, William K. Wimsatt, and William Empson, for example, considered poetic figuration as 'manifestations,

[1] Seamus Perry, *Coleridge and the Uses of Division*, Oxford English Monographs (Clarendon Press, 1999), p. 79.

[2] Marjorie Levinson, *Thinking Through Poetry: Field Reports on Romantic Lyric* (Oxford University Press, 2018), p. 19.

[3] Dries Vrijders, 'History, Poetry, and the Footnote: Cleanth Brooks and Kenneth Burke on Keats's "Ode on a Grecian Urn"', *New Literary History*, 42.3 (2011), pp. 537–552 (p. 541).

in language, of a fundamental unity that encompasses both mind and object'.[4] Defining organicism in terms of poetic form, they saw it as an ideal structure that resolves all paradoxes and seeming incongruities in a literary work into a formal unity that reveals universal truths about nature and humankind.[5] As Paul de Man notes, this understanding of organicism has troubling ontological implications because it insinuates that nature is a set of signs that point beyond the phenomenal world to transcendental truths that dissolve difference into identity. This interpretation resembles Samuel Taylor Coleridge's theory of life in which the pole of individuation ultimately gives way to the undifferentiated essence of life. Just as Coleridge tried to abstract life from its physical instantiation (but ultimately failed), the New Critics turned away from the 'poem's individual historical content' to concentrate on 'transhistorical form', a move that relies on a misunderstanding of Romantic organicism.[6]

De Man criticizes mid-twentieth-century scholars who took Coleridge's formulation of subject–object relations as Romanticism's core problematic. In *Blindness and Insight* (1971, 1983), de Man dismantles organicism by deconstructing the symbol, which is organicism's literary expression. As a rhetorical figure, according to de Man, the symbol misrepresents the human condition by bolstering the misconception that the human subject is free from temporal constraints. He claims that in contrast to allegory and irony, which are structurally constituted by temporal displacement, the symbol conflates representation with reality thus locating meaning in an eternal, transcendental sphere that exists separately from the timebound phenomenal world. De Man contends that allegory depends on temporal difference to function because its signs refer to concepts that are already in circulation. For example, the City of Destruction in John Bunyan's *The Pilgrim's Progress* (1678) refers to the earlier concept of a sinful, postlapsarian reality, and the allegory obtains only because the figure relies on an extant Christian worldview. Irony relies on a different kind of displacement in which words signify a meaning that differs from their literal interpretation; without the disjuncture between representation and meaning, irony fails. Where the 'symbol postulates the possibility of an identity or identification, allegory designates a distance in relation to its own origin, and', de Man explains, 'prevents the self from an illusory identification with the non-self'.[7] By figuratively collapsing the diachronic and synchronic orders, the Romantic symbol enacts a 'defensive strategy that tries to hide from this negative self-knowledge' that human beings are not only timebound

[4] Paul de Man, *Blindness and Insight: Essays in the Rhetoric of Contemporary Criticism*, 2nd edn (Routledge, 2013), p. 194.

[5] This New Critical criterion excluded many excellent poems from the Romantic canon because they were not considered sufficiently literary.

[6] Vrijders, 'History, Poetry, and the Footnote', p. 540.

[7] De Man, *Blindness and Insight*, p. 207.

but also separate from each other and the world in which they live.[8] While de Man argues that the symbol and, by extension, its ideological substrate organicism offer an inauthentic refuge against the existential human condition, he goes even further and considers the New Critics' dialectic interpretation of organicism as a fundamental mischaracterization of Romanticism.

A harbinger of the European post-structuralist and deconstructionist theories that reached a zenith in the 1980s and 1990s, de Man's work takes not just the New Critics but also mid-twentieth-century scholars such as Earl R. Wasserman and M.H. Abrams to task for identifying the dialectic between the subject (mind) and the object (nature) as the central problem of Romanticism. Where de Man considers William Wordsworth's and Jean-Jacques Rousseau's use of allegory and Friedrich Schlegel's theory of irony, Abrams and Wasserman hold Coleridge's 'dialectic of subject and object to be an authentic pattern of romantic imagery', thus shifting critical focus away from questions of form and structure towards issues of ontological priority, in 'an act of ontological bad faith'.[9] In so doing, they cast the Romantic poets as radical idealists or, in some cases, as dualists. The latter, as I argue in the following chapters, is particularly egregious given the Romantic objection to Cartesianism. These scholars' fixation with the subject–object dialectic, according to de Man, 'forces them ... into a persistent contradiction' that, in my view, leads them to distort intellectual history by mischaracterizing organicism.[10] For example, Thomas McFarland and Kathleen Coburn, who openly despise British men of science such as Erasmus Darwin and David Hartley, dismiss the influence of empiricism on Coleridge's development and describe his interest in them as a 'brief flirtation' that he quickly abandoned in favour of German metaphysics.[11] Likewise, Wordsworth scholars Alan Grob and Thomas Keith participate and, consequently, perpetuate early debates between Arthur Beatty and Melvin Rader about whether Wordsworth should be considered a Hartleyan or a transcendentalist. The trend, at least until the rise of New Historicism, has been to trace his philosophical development from an early investment in materialism to a 'Christian dualism of heaven and earth, soul and body, immortality and mortality', which he adopted sometime after 1805.[12] Mid-century scholars also had trouble interpreting these poets' work. For example, Coleridge's poem 'The Eolian Harp' stymied early-to-mid-twentieth-century scholars, especially when they tried to read it as an expression of his ontology, a topic that I take up in Chapter 3.

[8] De Man, *Blindness and Insight*, p. 208.

[9] De Man, *Blindness and Insight*, pp. 197, 211.

[10] De Man, *Blindness and Insight*, p. 197.

[11] Kathleen Coburn, *Inquiring Spirit: A Coleridge Reader* (Minerva Press, 1951), p. 27.

[12] Alan Grob, *The Philosophic Mind: A Study of Wordsworth's Poetry and Thought 1797–1805* (Ohio State University Press, 1973), p. 251.

While I agree with de Man's contention that the symbol has no special relationship to reality and offers no loophole for escaping humankind's existential condition and with his claim that the almost exclusive focus on subject-object relations functions as a red herring in Romantic scholarship, his turn to irony and allegory uncritically accepts the definition of organicism advanced by the scholars he critiques. Reframing the 'rediscovery of allegory' as the 'essential movement of Romanticism' and redefining its central problematic as the tension between the 'conception of the self seen in its authentically temporal predicament' and the 'impossible ambitions of the symbolic', de Man perpetuates their mistake.[13] In his view, the symbol signals the Romantics' desire for unmediated contact with reality, free from the distortions of language. Put another way, de Man claims that Romanticism centres on the conflict between existential separation and the strategies of containment that its authors employed to obfuscate this reality.[14] This definition, which New Historicists seized upon in the 1980s, is too narrow. As I argue below, British definitions of organicism point to the power of language to shape people's phenomenological perceptions and embodied experiences of the world.

In rejecting the 'teleological formalism associated with the old historicism, the dominant form in nineteenth-century historiography', New Historicists expanded de Man's problematic.[15] Where he saw Romantic poetry as an expression of tension between reality and the desire to escape temporal displacement, critics such as Levinson considered 'representation … always and by definition misrepresentation'.[16] Although Levinson contends that New Historicists did not accuse the poets of deliberately 'erasing the compromising evidence of contemporary life and politics', they nonetheless reframed Romanticism in terms of ideological obfuscation.[17] In *The Romantic Ideology* (1983), Jerome McGann states that Romantic poetry 'presents an image of an organic society and processive history … because such an ideology helps to maintain the status quo and to conceal the truth about social relations: that the rich and the ruling classes dominate the poor and the exploited'.[18] In the wake of McGann's epoch-making argument, scholars began reading Romantic poetry with a hermeneutics of suspicion that rejected the concepts valued by critics

[13] Jean-Pierre Mileur, 'Allegory and Irony: "The Rhetoric of Temporality" Re-Examined', *Comparative Literature*, 38.4 (1986), pp. 329–336 (p. 330), doi:10.2307/1770393; de Man, *Blindness and Insight*, p. 208; Mileur, 'Allegory and Irony', p. 331.

[14] See Fredric Jameson, *The Political Unconscious: Narrative as a Socially Symbolic Act* (Cornell University Press, 1983).

[15] Levinson, *Thinking Through Poetry*, p. 34.

[16] Levinson, *Thinking Through Poetry*, p. 13.

[17] Levinson, *Thinking Through Poetry*, p. 13.

[18] Jerome McGann, *The Romantic Ideology: A Critical Investigation* (University of Chicago Press, 1983), p. 8.

in the first half of the twentieth century.[19] However, like de Man, these critics relied on established definitions of organicism, which has long been conflated with German transcendentalism. McGann clearly links the Romantic ideology with the 'German Ideology', that is with the 'displacement which [Karl] Marx saw as a drug on human consciousness' and which Marx and Friedrich Engel associated with the Kantian and neo-Kantian philosophies that emerged in the eighteenth and early nineteenth centuries.[20] In short, the two leading schools of criticism that challenged Romantic aesthetics, particularly its formulation as an ontological attempt to reconcile the external object with the human subject, uncritically accepted definitions of organicism advanced by scholars in the first half of the twentieth century. Despite their many contributions to the study of Romantic literature and culture, they left the basic definition of organicism uncontested.

Recently, scholars such as Joan Steigerwald and Jennifer Mensch, in *Experimenting at the Boundaries of Life* (2019) and *Kant's Organicism* (2013) respectively, have returned German organicism to its roots in mid-to-late-eighteenth-century experiments and debates about epigenesis. While neither Steigerwald nor Mensch argues against the transcendentalist turn, both show how it was predicated on material investigations into bodies. Similarly, this monograph makes an intervention into British organicism by tracing it back to Hartley's and Darwin's medical fascination with embodiment and resituating it in the contexts that shaped its development as a theory—resistance to Cartesianism, the burgeoning science of the mind, and debates about the source of life. Like Denise Gigante, in *Life* (2009), I argue that returning organicism to its historical scientific origins requires us to understand Romantic aesthetics in new ways. However, where she looks at monstrous literary forms, I focus on embodied poetics.

I argue that British formulations of organicism do not obscure the realities of class exploitation. Rather, they try to expose social inequity and offer an alternative way of envisioning and experiencing one's relationship to society and the natural world. For example, Darwin's account of life imagines a world grounded in organic reciprocity and emphasizes mutual co-dependency and relationality. Similarly, Wordsworth's organicism exposes the social problems caused by exploitive class relationships and offers a metaphorical model that can alter one's embodied experience of the world. Here, losing sight of the Cartesian context that spurred these theories of organicism makes it more difficult to see how they sought to make linguistic and embodied interventions into current social, political, and economic conditions.

[19] Paul Ricoeur, *Freud and Philosophy: An Essay on Interpretation* (Yale University Press, 1970).

[20] McGann, *The Romantic Ideology*, p. 26.

Cartesian Dualism: Then and Now

Perhaps the most important ideological construct to emerge from the Enlightenment and Romantic eras is what N. Katherine Hayles calls the 'liberal humanist subject'.[21] Originating with seventeenth-century French mathematician and philosopher René Descartes, Cartesianism lays the groundwork for the concept of an autonomous human subject defined as an integrated self, unified by consciousness, and imbued with freewill. It radically separates the thinking mind from its physical substrates—both the body and the phenomenal world—asserting that they consist of two different substances. In his *Discourse on Method* (1637), Descartes attempts to find the 'true method of arriving at knowledge of everything the mind was capable of grasping' to guide scientific investigations, an approach that he employs in *Meditations on First Philosophy* (1641), which tries to prove God's existence.[22] Detaching his 'mind from the senses', he claims that he has 'so accurately observed that there are very few things one can know with certainty about corporeal objects, that there are many more things which are known to us about the human mind' that he can now 'turn [his] mind away without difficulty from the consideration of sensible or imaginable things, in order to bring it to bear on those which, being disengaged from all matter, are purely intelligible'.[23] From these meditations, Descartes asserts that sensory evidence is a less reliable form of knowledge than thought and, privileging mind over matter, concludes that the mind 'needs no place and depends on no material thing'.[24] According to his viewpoint, the mind is an immaterial substance that can exist 'independently of physical instantiation'.[25] Indeed, Descartes's assertion of the *cogito* marked a new epoch in Western philosophy that many Romantic thinkers from across Europe found objectionable.

Eighteenth-century French materialist Baron d'Holbach and English Dissenter and man of science Joseph Priestley, along with Coleridge, identified Descartes as the first thinker who separated 'that which thinks … from matter'.[26] In place of earlier ontological and epistemological paradigms that had

[21] N. Katherine Hayles, *How We Became Posthuman: Virtual Bodies in Cybernetics, Literature, and Informatics* (University of Chicago Press, 1999), p. 5.

[22] René Descartes, *Discourse on Method and the Meditations*, trans. by F.E. Sutcliffe (Penguin Books, 1971), p. 40.

[23] Descartes, *Method and Meditations*, p. 132. The full title of the 1641 work is *Meditations on the First Philosophy in which the Existence of God and the Real Distinction Between the Soul and the Body of Man are Demonstrated.*

[24] Descartes, *Method and Meditations*, p. 54.

[25] Ray Jackendoff, *Consciousness and the Computational Mind* (MIT Press, 1987), p. 15.

[26] Paul Henri Thiry d'Holbach, *The System of Nature; or, the Laws of the Moral and Physical World*, trans. by William Hodgson, 4 vols (G. Kearsley, 1797), I, pp. 166–167, Eighteenth Century Collections Online <link.gale.com/apps/doc/CW0122703088/ECCO?u=asecs&sid=bookmark-ECCO&xid=cf0c90a7&pg=1> [accessed 15 April 2012].

dominated since at least the Middle Ages, Cartesianism emptied the natural world of vitality and split human beings into body and soul. Although the Christian Schoolmen believed the soul was a 'spiritual substance', they did not wholly separate it from the body, and the Church supported this view, 'condemn[ing] the opinion that the intellectual soul was not the substantial form of the body'.[27] However, once Cartesian dualism was widely accepted, the 'distinction between the soul and the body took on a more thoroughly literal rather than metaphorical meaning', which I contend altered the way people understood and experienced their *umwelt*.[28] Priestley succinctly summarizes the Cartesian position, stating that 'Matter is that kind of substance of which our bodies are composed, whereas the principle of perception and thought belonging to us is said to reside in a spirit, or immaterial principle', a notion that he finds reprehensible, not least because it vilifies the body.[29] Substance dualism classifies the body and the phenomenal world as material substances, which Cartesian and Newtonian physics characterize as passive, inert, and lifeless. In contrast, the immaterial mind or soul is the active principle that is responsible for all movement, intellection, and motive.[30] According to this system, immateriality connects humankind to God, also an immaterial substance, and while this accounts for the so-called divine spark in human beings, it does so at the expense of the rest of creation.

Although Descartes held that humans are made up of both the body and the soul, he nonetheless maintained a sharp distinction between them, which raises the vexed question of interactionism. That is, how does the material body transmit sense data to the immaterial soul, and, vice versa, how does the soul make the body act? Despite his attempt to link the soul to the body through the pineal gland, Descartes recognized that he did not formulate a satisfactory 'explanation of the union between the soul and the body, which', he admits, 'I have not yet dealt with at all'.[31] Thus, in separating the mind from the body and the world, substance dualism creates an 'explanatory gap between mind and matter, consciousness and nature'.[32] The separation of the

[27] Brother Azarias, *Aristotle and the Christian Church* (Kegan Paul, Trench, and Co., 1888), pp. 108–109.

[28] Thomas Dixon, *From Passions to Emotions: The Creation of a Secular Psychological Category* (Cambridge University Press, 2003), p. 76.

[29] Joseph Priestley, *Disquisitions Relating to Matter and Spirit* (Kessinger Publishing, 2005), p. xxviii.

[30] The terms mind, soul, and spirit were used interchangeably in the mind–matter debates.

[31] René Descartes, *The Philosophical Writings of Descartes*, trans. by John Cottingham, Robert Stoothoff, and Dugald Murdoch, 2 vols (Cambridge University Press, 1984), II, p. 275.

[32] Evan Thompson, *Mind in Life: Biology, Phenomenology, and the Science of Mind* (Harvard University Press, 2007), p. 6.

immortal soul from a perishable substance also appealed to Christians such as Cambridge Platonist Ralph Cudworth, who, by 'rejecting these attempts to link thought with matter, … believed he was undermining the grounds for atheism'. He nonetheless could not explain how these two very different substances interact.[33]

Later in the seventeenth century when John Locke published *An Essay Concerning Human Understanding* (1690), he exacerbated the problems surrounding the mind–matter debates. Going through four editions in just sixteen years, all of which included substantive additions, Locke's *Essay* responds to Descartes's *Meditations* with the claim that infants are not born with innate ideas but come into the world with minds like 'white paper, void of all characters, without any ideas'.[34] The 'great source of most of the ideas we have', Locke contends, depend 'wholly upon our senses, and [are] derived by them to the understanding'.[35] Although he staunchly embraced dualism, his epistemology undermines it by making cognition dependent on the sensate body. Thus, even as Locke accepted Descartes's ontology, his quarrel with Cartesian epistemology intensified the question of interactionism.

These ontological and epistemology conundrums survived into the eighteenth century, which saw an increase in scientific interest in the body. Not only did thinkers, mostly medical men, want to understand how it worked, but it was becoming increasingly clear that neither Cartesian nor Newtonian physics was adequate to explain living beings. By setting up a binary opposition between insensible matter and an active immaterial substance, Cartesianism turns nature into a system of strings, pullies, and levers that cannot move without being set into motion. From a phenomenological perspective, the mechanistic view of nature contradicts many people's experiences of it. For example, in Descartes animals are 'beast-machines', who are driven solely by instinct.[36] The implication is that they cannot suffer because they lack active intelligence. While physicians were trying to understand cognition as a physiological process, 'life scientists', explains Gigante, 'focused on the dynamics of organic form in an effort to explain how form emerged and maintained itself, despite the physical laws' that regulate inanimate nature.[37] As they began

[33] John W. Yolton, *Thinking Matter: Materialism in Eighteenth-Century Britain* (University of Minnesota Press, 1983), p. 8.

[34] John Locke, *An Essay Concerning Human Understanding*, 4th edn (Prometheus Books, 1995), p. 59.

[35] Locke, *Human Understanding*, p. 60. Ideas of reflection arise when the mind contemplates its own action.

[36] C.U.M. Smith, 'Julien Offray de La Mettrie (1709–1751)', *Journal of the History of the Neurosciences*, 11.2 (2002), pp. 110–124 (p. 115), doi:10.1076/jhin.11.2.110.15188.

[37] Denise Gigante, *Life: Organic Form and Romanticism* (Yale University Press, 2009), p. 5.

examining the structures of living beings to determine how bodies grow from seedlings or embryos into mature organisms, they 'concluded that there must be some "productive power" in nature that enabled unorganized material to generate new living forms', that is, the lifeforce.[38] Surgeons John Hunter and John Abernethy argued that this lifeforce was something like the divine spark that, rather than being an innate part of the body, was superadded by God. Organicists such as Darwin and Coleridge, albeit in very different ways, tended to see life as an emergent property that organized living bodies from within. Their theories of organicism not only address the relationship between mind–body–world but see the lifeforce as a driving principle. Thus, these debates about cognition and life provide important contexts both for understanding Romantic organicism and for understanding early-to-mid-twentieth-century critics' investment in it.

Perry asserts that the 'philosophical heart of Romanticism' consists of a 'prolonged and strenuous attack on the Cartesian view'.[39] In decoupling the human subject from the object-world in which it lives, Cartesian dualism rests on a metaphor of isolation and separation. As a structuring concept, it implicitly sanctions self-interested actions that either explicitly or inadvertently cause harm to others and to the environment by positing the phenomenal world, which includes the human body, as impermanent and unimportant. Because many Romantic thinkers recognized the ominous underpinnings of substance dualism, it became a major intellectual problem in the eighteenth, nineteenth, and even twentieth, centuries. The question became more contentious in light of new theories of cognition and life. Thinkers wrestled with whether to situate the development of humankind in anthropology or natural history. To maintain human evolution as separate from the development of the natural world (anthropology) risks perpetuating Cartesian dichotomies, while situating it in nature (natural history) blurs the boundaries between humankind and other natural objects. As a compromise, many organicists tried to posit non-Cartesian paradigms of organicism to reconcile the ostensibly unique human capacity for consciousness, which they associated with the ability to reason and the freedom of will.

For example, German *Naturphilosophen* advanced models of organicism that combined ontology, metaphysics, and speculative science in an attempt to resolve the separation that Cartesian dualism had imposed on the mind, body, and nature, but they could not disencumber themselves of the desire for a free and autonomous human subject. As Michael Rosen notes, German idealism centres on a 'single fundamental concept … freedom', which thinkers from 'Kant to Hegel' explored in response to theological and scientific questions

[38] Gigante, *Life*, p. 12.

[39] Perry, *Coleridge and Division*, p. 122.

about the 'situation of the agent in a deterministic universe'.[40] In attempting to ground freedom in nature, philosophers such as Friedrich Wilhelm Joseph von Schelling fell prey to the 'temptation' identified by de Man 'for the self to borrow … the temporal stability that it lacks from nature'.[41] In other words, Romantic transcendentalists tried to overcome dualism while preserving the Cartesian subject because, as de Man contends, they feared the existential alternative.

However, in this monograph I bring to light non-transcendentalist theories of Romantic organicism, arguing that they are crucial precisely because inimical Cartesian modes of envisioning the mind–world relationship persisted into the twentieth and twenty-first centuries. This persistence accounts for early scholars' attachment to the Cartesian subject and explains why they both defended transcendentalist theories of organicism and fixated on questions of consciousness and the nature of subject–object relations. Adding to de Man's suggestion that the urge to find temporal stability for the human subject is 'present, though perhaps not consciously, in critics such as Abrams and Wasserman', I claim the same applies to the scholars that predate them.[42] Their need to protect the sanctity of the autonomous subject reveals a deeper need to stave off the existential panic induced by threats to Cartesian subjectivity in psychology and cognitive science.

In the early twentieth century, Sigmund Freud's psychoanalytic theory divided the human subject into the ego and id, both of which are rooted in the unconscious. Speculating that 'repression is a normal part of human development', Freud not only undermines the notion that consciousness drives subjectivity, but he makes psychological health dependent on aspects of mind that are out of reach of our awareness.[43] Later, in the middle of the century, the rise of cybernetics posed further threats to the unified subject by reframing cognition in terms of information processing systems, feedback loops, and a 'flow of binary code pulsing through neurons'.[44] Viewing the 'human being as an informational pattern', cyberneticists such as Norbert Wiener and Claude Shannon weakened the boundaries between humanity and machines.[45] However, they resisted the implications of their research and its potential, as Hayles argues, to undermine the liberal ideal of 'a coherent, rational self, the

[40] Michael Rosen, 'Freedom in German Idealism', *Harvard University*, May 2007, pp. 1, 3 <https://scholar.harvard.edu/michaelrosen/publications/freedom-german-idealism> [accessed 11 March 2025].

[41] De Man, *Blindness and Insight*, p. 197.

[42] De Man, *Blindness and Insight*, p. 197.

[43] Dino Franco Felluga, *Critical Theory: The Key Concepts*, Routledge Key Guides (Routledge, 2015), p. 267.

[44] Hayles, *How We Became Posthuman*, p. 61.

[45] Hayles, *How We Became Posthuman*, p. 61.

right of that self to autonomy and freedom, and a sense of agency linked with a belief in enlightened self-interest'.[46] In contrast to Romantic men of science Hartley and Darwin, who happily blurred the boundaries between humans and animals, Weiner worked to maintain the fiction of an autonomous self for 'fear that if the boundaries are breached at all, there will be nothing to stop the self's complete dissolution'.[47] Considering the cyberneticist response, it is unsurprising that Romantic scholars clung to a concept of organicism that protects the idea of a transcendental self that rises above the vagaries of temporal materiality. Unfortunately, as scholars such as Saree Makdisi have argued, the liberal humanist subject is ultimately normative and regulatory, offering a vision of political freedom based on essentializing differences.[48]

Positing a subject that is unique because it is different from everything else in the natural world, Cartesian dualism conceptually disconnects humankind from the environment in which it is embedded and places the cogitating human above non-thinking nature. Aside from the ontological problems this raises, it disrupts the phenomenological sense of connection that people might feel with their surroundings and their communities. The implications of this metaphorical separation have real life consequences, both then and now. In addition to authorizing humans to exploit nature without considering the long-term ramifications, it also sanctions the exclusion of living beings deemed to be below a certain cognitive threshold from the category of human, including 'othered' people. According to Stefani Engelstein, 'only in the context of an increasingly mechanized world could physicians, surgeons, and naturalists begin to recognize the regularity and reciprocal activity of the parts of the living body as rule-governed'.[49] Thus, Cartesianism not only paved the way for organicism but also created its conditions of possibility by enabling new conceptions of the body as a system. Nonetheless, I argue that divesting ourselves of our attachment to Cartesian subjectivity, as twenty-first-century cognitive science invites us to do, and recollecting that Romanticism was largely an anti-Cartesian project, enables us to see organicism's potential.

Organicism: A Brief Overview

In *Romantic Organicism: From Idealist Origins to Ambivalent Afterlife* (2003), Charles I. Armstrong reminds readers that organicism functions as a 'grounding

[46] Hayles, *How We Became Posthuman*, pp. 85–86.

[47] Hayles, *How We Became Posthuman*, p. 290.

[48] See Saree Makdisi, 'Blake and Romantic Imperialism', in *William Blake and the Impossible History of the 1790s* (University of Chicago Press, 2003), pp. 204–259.

[49] Stefani Engelstein, *Anxious Anatomy: The Conception of the Human Form in Literary and Naturalist Discourse*, Studies in the Long Nineteenth Century (State University of New York Press, 2008), p. 156.

systematics for understanding all holistic structures'.[50] As a response to Cartesianism, he suggests that organicism tries to theorize dynamic, integrated systems, 'first, to demonstrate that nature cannot be explained mechanistically as a series of causes and effects; and second, to undermine the dualistic standpoint'.[51] Attempting to repair the unbreachable division between matter and spirit, organicists attempt to reintegrate the natural and the human, while accounting not only for the common lifeforce that runs throughout nature but also for intelligence, both human and nonhuman. Thinkers across Europe developed models of organicism in response to the emerging life sciences. They tried to envision the world as an interdependent system that developed like a living organism. Although the word 'organism' was not in common use until the 1830s, Immanuel Kant's notion of a 'natural end' served as its conceptual equivalent and provided a model of an organic system, particularly in Germany.[52]

Seeking to establish a clear boundary between living and non-living systems, Kant theorizes organisms as systems in which the parts and the whole exist in a mutually co-productive, reciprocal relationship—a necessary condition for the production and maintenance of life. In *Critique of the Power of Judgement* (1790), he explains that in a

> product of nature each part is conceived as if it exists only through all the others, thus as if existing for the sake of the others and on account of the whole, i.e., as an instrument (organ), which is, however, not sufficient, ... rather it must be thought of as an organ that produces other parts (consequently each produces the others reciprocally), ...only then and on that account can such a product, as an organized and self-organizing being, be called a natural end.[53]

In Kant's terminology, 'a product of nature' refers to any object that has not been created by human means, while 'a natural end' refers to a living being. To

[50] Charles I. Armstrong, *Romantic Organicism: From Idealist Origins to Ambivalent Afterlife* (Palgrave Macmillan, 2003), p. 2.

[51] Devin Zane Shaw, 'The "Keystone" of the System: Schelling's Philosophy of Art', in *The Palgrave Handbook of German Idealism*, ed. by Matthew C. Altman, Palgrave Handbooks (Palgrave Macmillan, 2014), pp. 518–538 (p. 529).

[52] Gottfried Wilhelm Leibniz introduced the word 'organisme' in the early seventeenth century in French, but it was not until the 1830s that it began to be widely used in British discourse. See Charles T. Wolfe, 'Determinism/Spinozism in the Radical Enlightenment: The Cases of Anthony Collins and Denis Diderot', in *Boundaries in the Eighteenth Century*, ed. by Pasi Ihalainen and others (Finnish Society for Eighteenth-Century Studies, 2007), pp. 37–51 (p. 48).

[53] Immanuel Kant, *Critique of the Power of Judgment*, ed. by Paul Guyer, trans. by Eric Matthews (Cambridge University Press, 2000), sec. 5, p. 374.

be considered living, the constituent parts of the 'product' must be connected in such a way that they can only exist in relationship to each other. Additionally, all the parts must work towards the same 'end', that is, the production and maintenance of itself as a living organism. Indeed, even if a given system's parts form a functioning system, this is 'not sufficient', according to Kant, to be considered organic. In contrast to mechanical systems, the parts must not only work together, but they must also co-produce each other. They must be defined in relation to each other and in relation to the whole being that they maintain. In other words, an organism's life processes—digestion, cell reproduction, and so forth—bring it into being and keep it alive. At the same time, the whole depends equally on its parts not just for its identity but for its existence.[54]

Although the concept of organic form has been in circulation for centuries, Armstrong contends that thinkers in the Romantic period theorized it more radically as a 'structure that does not merely co-ordinate form and matter (or subject and object) but tries to collapse opposites into an immediate and undifferentiated unity'.[55] While the adjective 'undifferentiated' is inaccurate, in my view, the endeavour to unify form and matter into a single system captures what many thinkers, including those not typically considered organicists, attempted. Under Armstrong's definition, French and British philosophies of mind and life that have typically been considered mechanistic qualify as models of Romantic organicism. French *philosophes* such as Denis Diderot, Baron d'Holbach, and Julien Offray de La Mettrie, who were hostile towards their countryman Descartes, also reconceptualized the relationship between nature and human beings to close the Cartesian gap. They did this by theorizing the human body as a system that they situated into a larger system, nature, integrating humankind into natural history. In Britain, Darwin, Priestley, and Hartley did the same, advancing organic models of mind and life that envisioned them as part of a system of nature. Despite arguments by scholars who see British empiricism as the 'dead weight from which Wordsworth and Coleridge had to release themselves before they could take flight on the wings of the Kantian transcendental', I argue that Hartley's and Darwin's theories offer alternatives to Kant's anthropological approach and that they inflected the poets' thinking about organicism.[56]

[54] For an excellent primer in German Romantic philosophy, see *The Palgrave Handbook of German Idealism*, ed. by Matthew C. Altman, Palgrave Handbooks (Palgrave Macmillan, 2014).

[55] Armstrong, *Romantic Organicism*, p. 4.

[56] Craig Cairns, '"Kant Has Not Answered Hume": Empiricism and the Romantic Imagination', in *Romantic Empiricism: Poetics and the Philosophy of Common Sense, 1780–1830*, ed. by Gavin Budge, The Bucknell Studies in Eighteenth-Century Literature and Culture (Bucknell University Press, 2007), pp. 40–63 (p. 43).

My recovery of British organicism as a literary and cultural concept for the Romantic period and today serves as a corrective to the accounts advanced by the likes of Abrams, who attributes the 'first clear channel for the flow of organicism into the hitherto clear, if perhaps not very deep, stream of English aesthetics' to Coleridge's interest in German metaphysics.[57] By drawing attention to the central role of the body in key texts by Hartley, Darwin, Coleridge, and Wordsworth, I challenge the critical commonplace that defines organicism in terms of ideal types and metaphysics. In attending to the British context, this monograph complements recent studies in German organicism that argue for the importance of the body not only to *Naturphilosophie* but also to Kant's transcendental philosophy. Steigerwald reminds readers that Albrecht von Haller's and Johann Friedrich Blumenbach's experiments to better understand irritability, sensibility, and reproduction in the mid-eighteenth century were 'an important starting point … for the exploration of organic vitality in the years around 1800'.[58] In other words, German transcendentalist theories originate in studies of the body, according to Steigerwald and Mensch, even if the scholarly 'focus on organicism has drawn attention to ideal types and comparative anatomy … often to the neglect of studies of dynamic functions' of living creatures.[59] Even Kant's critical philosophy, including the idea of transcendental freedom, Mensch argues, depends on his understanding of epigenesis.[60] Engelstein's contention that eighteenth-century physiology provoked anxiety about the body by dismantling its physiological integrity and reducing it to a system of organs further attests to its centrality to organicism by virtue of its widespread cultural importance.[61] Similarly, revisiting British scientific accounts of embodiment underscores the body's crucial role in organicism.

Returning organicism to its basis in embodiment has significant implications for Romantic aesthetics. Rather than viewing moments in which the subject 'leaves the body and its brain behind' and 'succeeds in touching on the noumenal itself', theorists such as Hartley and Darwin reframe transcendent, sublime, and other ecstatic encounters as deeply embodied experiences.[62] Aesthetic encounters have physiological ramifications with long-lasting effects

[57] M.H. Abrams, *The Mirror and the Lamp: Romantic Theory and the Critical Tradition* (Oxford University Press, 1971), p. 168.

[58] Joan Steigerwald, *Experimenting at the Boundaries of Life: Organic Vitality in Germany around 1800* (University of Pittsburgh Press, 2019), p. 41.

[59] Steigerwald, *Experimenting at the Boundaries of Life*, p. 14.

[60] Jennifer Mensch, *Kant's Organicism: Epigenesis and the Development of Critical Philosophy* (University of Chicago Press, 2013), p. 15.

[61] Engelstein, *Anxious Anatomy*, p. 113.

[62] Alan Richardson, *The Neural Sublime: Cognitive Theories and Romantic Texts* (Johns Hopkins University Press, 2010), p. 24.

that remain in the body even if they do not always stay in the conscious memory. Hartley and Darwin placed these experiences at the core of their cognitive theories, seeing them as integral to human development with respect to character formation, knowledge production, and social formation. Additionally, the idea that aesthetic encounters produce embodied rather than disembodied experiences also play a key role in Coleridge's and Wordsworth's poetic theories. While discussions about organic aesthetics often focus on the relationship between the fragment and the whole or on art's ability to express transcendent truths, I argue that Romantic organicism and its attendant aesthetics share an ethical imperative to get readers to view and experience the world as a web of relations.

Aesthetics and Ethics

The New Critical definition of organicism mirrors the aesthetic theory that emerged from German idealism. The idealist belief that poetic language transcends ordinary language stemmed from the German axiom that 'poetry is … autonomous and submits to no law external to itself'.[63] Bringing together investigations into the lifeforce with aesthetics, neo-Kantians such as Novalis and Johann Wolfgang von Goethe viewed experimental science and art as 'figurative modes [that] do the work of mediation' between the phenomenal and the transcendental.[64] Trying to reconcile the subject and object realms, they saw art as a way to render 'thinking more concrete and seeing more abstract'.[65] For example, Schelling believed that 'what philosophy constructs in the ideal, art produces in the real', making tangible philosophical 'truths'.[66] Devin Zane Shaw explains that both the 'natural organism and the artwork embody the same identity of real and ideal', perfectly uniting the objective and the subjective, but 'the work of art overcomes these oppositions through the identity of conscious and unconscious production, whereas the organism's activity is unconscious'.[67] In other words, organisms self-realize through life processes that occur below the surface of conscious awareness, while human beings have the cognitive capacity for self-consciousness, which enables them to conceptualize themselves as both subjects and objects. This awareness allows them to consciously channel the unconscious productive drive of an artwork, helping it to take shape via the power of the creative imagination. In Schelling's view, the synthesizing faculty grants human beings a kind of freedom from

[63] Steigerwald, *Experimenting at the Boundaries of Life*, p. 233.
[64] Steigerwald, *Experimenting at the Boundaries of Life*, p. 395.
[65] Steigerwald, *Experimenting at the Boundaries of Life*, p. 26.
[66] Shaw, 'The "Keystone" of the System', p. 519.
[67] Shaw, 'The "Keystone" of the System', p. 519.

deterministic laws that other living beings lack; thus, art represents human freedom not only because it mediates the real and the ideal, but also because it is produced through artists' conscious and unconscious activity.

Schelling's aesthetics distinguishes human beings from organisms on the basis of consciousness, by which philosophers really mean self-consciousness. 'Consciousness', George di Giovanni explains, is the 'point where organic nature acquired its highest point of concentration by reflecting upon itself as nature and thus became spirit'.[68] Here, the ability to understand oneself is a natural end that proves human spirituality. Consciousness grounds the unconditional freedom that transcendentalists claim characterizes the human condition. With respect to aesthetics, the imagination is the apotheosis of the real and ideal because it is both physiological and spiritual. As New Historicists have noted, this understanding of aesthetics presents itself and art as free from timebound factors such as politics or culture. Art represents the fantasy that de Man identifies, the existence of an autonomous self that is free from temporality. Attending to the body in aesthetic theory thus disrupts this understanding of Romantic aesthetics.

Embodiment offered British thinkers a way to reconcile the 'fissure between a timeless but expansive … sympathetic sense of *nature*' represented in much of the poetry with a 'timebound, practical … sense of *history*' that grounds aesthetic experiences.[69] Romantic theories of embodied cognition provide the key for understanding such experiences. The figures discussed below did not require *a priori* transcendental categories of mind to account for mental and emotional phenomena because they attributed them to physiological structures. Thus, they viewed emotional responses to aesthetic experiences as phenomenological: although percipients might feel as if they are transcending physicality, the experience is, in fact, deeply embodied. When recognized as such, these experiences can be directed towards cultivating certain kinds of knowledge about the world. For example, in *The Prelude* (1799, 1805, 1850) and *The Excursion* (1814), Wordsworth nudges readers towards an anti-Cartesian experience of the world by representing ecstatic experiences as fundamentally relational by grounding them in interactions with the phenomenal world.

In both Wordsworth's and Coleridge's poetry, the feeling of transcendence alone does not make sublime or beautiful experiences important, but rather the opportunity for reflection it enables. Although Wordsworth defines poetry as the 'spontaneous overflow of powerful feelings', he crucially states that these

[68] George di Giovanni, 'Kant's Critical Legacy: Fichte's Constructivism and Hegel's Discursive Logic', in *The Palgrave Handbook of German Idealism*, ed. by Matthew C. Altman, Palgrave Handbooks (Palgrave Macmillan, 2014), pp. 665–686 (p. 673).

[69] Glenn Willmott, 'Blood in the Blue Lagoon: Byron and the Politics of Wonder', in *Awake to Love and Beauty: Proceedings from a Conference in Honour of George Whally*, ed. by Michael John DiSanto and others (Algoma University, 2016), pp. 65–75 (p. 65).

'emotion[s must be] recollected in tranquillity'.[70] According to Nancy Yousef, his aesthetics 'compel a reflective pause over the tendency, and the desire, to draw sharp distinctions between the moral and the animal, the thoughtful and emotional, the spiritual and the passionate, the poetic and the philosophical'.[71] Wordsworth invites readers to break with the binary thinking engendered by Cartesian dualism and to think relationally instead. The physiological effects of poetry, art, and nature account for the transformative experiences that Romantic thinkers associated with aesthetics. For Hartley, the feelings from encounters with the sublime and the beautiful replicate those generated by communion with the divine and encourage people to grow closer to God. For Darwin, the beautiful is fundamental to the formation of society, while Coleridge's poetry offers readers an opportunity to practise the method that he outlines in *The Friend* (1818). In all these cases, aesthetics does not work to resolve contraries to achieve unity. Rather, it becomes an opportunity to reflect on embodied experiences and to cultivate a sense of relationship with one's surround, which forms the basis of an ethical aesthetics.

Despite Wordsworth and Coleridge's self-proclaimed goal to break with tradition and pioneer new forms of poetry, Romantic aesthetics was rooted in the early eighteenth-century moral philosophy of Scottish philosophers such as Lord Shaftesbury and Francis Hutcheson, who argued that aesthetics and morality are inextricably linked. Shaftesbury claimed that human beings have an innate moral sense that enables them to discern right from wrong and that naturally impels them towards virtue, just as surely as they recognize beauty and gravitate towards it. He saw moral sense 'in its natural condition, [as] wholly or mainly emotional'.[72] Responding to Hobbesian views that human beings are naturally self-serving, in his *Characteristics of Men* Shaftesbury

[70] William Wordsworth, 'Preface' to *Lyrical Ballads*, *The Prose Works of William Wordsworth*, ed. by W.J.B. Owen and Jane Worthington Smyser, 3 vols (Clarendon Press, 1974), I, p. 148, ll. 387–388. Note that Wordsworth uses the word emotion, which Thomas Dixon argues does not correspond with the modern psychological definition that it has today. Despite Dixon's claim that the contemporary meaning, which came about in the mid-to-late nineteenth century, 'tends to include many feelings that previously have been categorised not as passions but as appetites (e.g. lust), or affections (e.g. religious feelings), or sentiments (e.g. sympathy)', the thinkers in this study, from Wordsworth to Hartley to Darwin to Coleridge, use the term in this more contemporary way as well as in the older sense. Dixon, *From Passions to Emotions*, p. 18. For example, Hartley uses the word emotion interchangeably with affection and passion throughout *Observations*.

[71] Nancy Yousef, 'After-Affects and Second Thoughts: Wordsworth, Eliot, and the Forms of Emotional Thinking', in *Romanticism and Consciousness, Revisited*, ed. by Richard C. Sha and Joel Faflak, Edinburgh Critical Studies in Romanticism (Edinburgh University Press, 2022), pp. 144–166 (p. 157).

[72] Thomas Fowler, *Shaftesbury and Hutcheson*, English Philosophers, 5 (G.P. Putnam's Sons, 1883), p. 83.

contends that just as people are drawn to the 'fair and shapely, the amiable and admirable' and repulsed by 'the deformed, the foul, the odious, or the despicable', they are also predisposed to act virtuously because doing so generates feelings akin to those produced by encounters with the beautiful.[73] Like Wordsworth after his encounter with Godwinian rationalism, Shaftesbury associates morality with feeling rather than reason, asserting that 'Wisdom is more from the *Heart*, than from the *Head*. *Feel* Goodness, and you will see all Things fair and good'.[74] Thus, while Shaftesbury acknowledges the benefits of reasoned reflection, he emphasizes the emotional underpinnings of moral and aesthetic responses. In a similar vein, Hartley and Darwin put emotion at the centre of their theories of cognition, laying the foundation for an ethical aesthetics based on human physiology.

Developing Shaftesbury's concept of moral sense, Hutcheson places more emphasis on pleasure and pain, which brings him closer to Hartley and Darwin, who based their theories of cognition on these two basic sensations. Like Shaftesbury, he claims that people recognize and react immediately to virtue and vice, but he thinks that it is because moral sense 'respond[s] to the moral quality perceived in actions (their benevolence or lack of it) and [leads] us to approve of agents or to dislike them according to the pleasure or pain accompanying perception'.[75] Moral sense for both Hutcheson and Shaftesbury functions as conscience, which evaluates the morality of actions, steering us towards the good and tormenting us with remorse when we act immorally. Hutcheson states that the conscience 'does not estimate the good it recommends as merely differing in degree … so as to allow us to practice smaller moral evils' but understands good and evil as absolutes.[76] Moral philosophy was crucial to the rise of sentimental fiction and novels of sensibility, which appealed to readers' emotions in an effort to get them to sympathize with socially or economically marginalized characters, their authors believing that it would improve their readership's moral sense and lead to benevolent action.[77]

73 Anthony Ashley Cooper Shaftesbury, *Characteristics of Men, Manners, Opinions, Times, Etc*, ed. by John Mackinnon Robertson, 2 vols (G. Richards, 1900), II, p. 137, Internet Archive <http://archive.org/details/characteristicso02shaf> [accessed 27 June 2024].

74 Anthony Ashley Cooper Earl of Shaftesbury, *Several Letters Written by a Noble Lord to a Young Man at the University* (J. Roberts, 1716), p. 32.

75 Daniel Carey, 'Francis Hutcheson's Philosophy and the Scottish Enlightenment', in *Scottish Philosophy in the Eighteenth Century: Morals, Politics, Art, Religion*, ed. by Aaron Garrett and James A. Harris, A History of Scottish Philosophy, 2 vols (Oxford University Press, 2015), I, pp. 36–76 (p. 40).

76 Francis Hutcheson, *A System of Moral Philosophy, in Three Books*, 3 vols (A. Millar, 1755), I, p. 61.

77 For discussions of sentimental fiction and its connection to moral philosophy, see Janet Todd, *Sensibility: An Introduction* (Methuen, 1986) and Maureen Harkin, 'Introduction', in *The Man of Feeling* (Broadview Press, 2005), pp. 9–38.

Furthermore, in addition to eliciting emotional responses, the characters in novels of sentiment such as Sarah Fielding's *The Adventures of David Simple* (1744) or Samuel Richardson's *The History of Sir Charles Grandison* (1753) model benevolent behaviour for readers to emulate. Although Wordsworth reacted against the artifice of end-stopped heroic couplets, the same moral philosophers who inspired the language of feeling influenced his aesthetic theory.

Like the moral philosophers and sentimental novelists, and unlike the general attitude today, the Romantics did not view aesthetic experiences as ancillary to everyday life, but rather as essential to the fabric of society. They did not see literature as merely mimetic but believed that its effects reached beyond the pages of the book and could change people's behaviours for the better. When the cult of sensibility was at its peak, Adam Smith published *The Theory of Moral Sentiments* (1759) in which he argues that ethical action depends on the imagination because it enables people to sympathetically identify with others' suffering. These theories received backlash from those who questioned whether making someone feel was enough to inspire social change, but my point is that Romantic authors had a precedent for connecting literature to morality and ethical action.[78] However, where the novelists of sensibility considered the experience of sympathy to be adequate to motivate change, the poets emphasized the need to reflect on affective responses.

Additionally, the rise of nerve theory in the first half of the eighteenth century gave thinkers a way to conceive aesthetics not just in terms of form but also in terms of neurophysiological effects. For example, Scottish physician George Cheyne's nerve theory in *The English Malady* (1733) contributed to the popularity of the literature of sensibility, and, later, drawing on 'his involvement with medical materialism', Edmund Burke locates the response to encounters with the sublime and the beautiful in the nervous system.[79] In *A Philosophical Enquiry into Our Ideas of the Sublime and the Beautiful* (1757, 1759), he claims that these two aesthetic experiences undergird the two most

[78] There are a host of other issues at play here, including the gendered nature of authorship in the eighteenth century, and my necessarily brief discussion of the literature of sentiment and sensibility does not do it justice. To read more about the reaction against this literary movement, see Todd, *Sensibility* and Catherine Ingrassia, 'Introduction', in *Anti-Pamela and Shamela* (Broadview Press, 2004).

[79] Cheyne and Richardson were close friends, which may also account for the connection. For a discussion of the relationship between nerve theory and the cult of sensibility, see G.J. Barker-Benfield, *The Culture of Sensibility: Sex and Society in Eighteenth-Century Britain* (University of Chicago Press, 1992). See also Aris Sarafianos, 'Pain, Labor, and the Sublime: Medical Gymnastics and Burke's Aesthetics', *Representations*, 91.1 (2005), pp. 58–83 (p. 58), doi:10.1525/rep.2005.91.1.58 and 'The Contractility of Burke's Sublime and Heterodoxies in Medicine and Art', *Journal of the History of Ideas*, 69.1 (2008), pp. 23–48, doi:10.1353/jhi.2008.0005.

fundamental human drives: self-preservation and reproduction. The beautiful, he contends, makes men crave the 'society of the *sexes*, which answers to the purposes of propagation', but they also desire '*general society*, which we have with men and with other animals' such as dogs.[80] In *Zoonomia* (1794), Darwin, too, argues that society emerges from encounters with the beautiful in infancy and early childhood. Aesthetics had widespread implications in eighteenth- and nineteenth-century society in a way that is unrecognizable today. For example, according to Noel Jackson, the 'aestheticization of political discourse of the French Revolution' resulted because people believed in the 'capacity of the writer to receive, process, and communicate sensation … [and] to exert pressure and influence on the public mind'.[81] Understanding Romantic aesthetics as embodied rather than transcendental makes it possible to understand the ethical impact attributed to art and literature.

Theory and Method

In *Wordsworth's Great Period Poems* (1986), Levinson explains the impetus behind New Historicism, cautioning that the 'prolific contraries' that animate 'Romantic poetry and criticism … are not our family of conflicts'.[82] Responding to the same critics with whom de Man took issue, she urges contemporary scholars to historicize their objects of study because Romantic aesthetics responded to a very different political and cultural moment from their own. However, in *Thinking Through Poetry* (2018), she revisits the old New Historicist problem but with a new methodological approach that she calls 'ontological materialism' and a 'sampling of so-called "post-classical" frameworks in the physical and life sciences', which allow her 'to set aside both classical and intuitive pictures of subjects and objects' to gain new insights into the greater Romantic lyric.[83] In this book, I respond to both Levinson's earlier and more recent call. In her study of Wordsworth, she identifies 'Romantic transcendence', a concept closely associated with organicism, as a 'bit of a white elephant' that we must 'refuse ... until such time as we can trace its source and explain its character'; only then can we 'share in the poets' ecstasies, or recover them in a meaningful form'.[84] Until recently, scholars have lacked an adequate critical apparatus to make sense of Romantic aesthetics on its own

[80] Edmund Burke, *A Philosophical Enquiry into the Sublime and Beautiful*, ed. by Paul Guyer, Oxford World's Classics (Oxford University Press, 2015), p. 34.

[81] Noel Jackson, *Science and Sensation in Romantic Poetry* (Cambridge University Press, 2008).

[82] Marjorie Levinson, *Wordsworth's Great Period Poems* (Cambridge University Press, 1986), p. 57.

[83] Levinson, *Thinking Through Poetry*, pp. 1, 19.

[84] Levinson, *Wordsworth's Great Period Poems*, p. 57.

terms, but two twenty-first-century developments have made it possible for me to recover Romantic organicism in a meaningful form. First, in accordance with Levinson's invitation to revisit old questions with new post-classical scientific paradigms, I draw on recent 4e theories of cognition (outlined below), which provide me with theoretical frameworks to interrogate British theories of mind and life in ways that were unavailable to previous generations of scholars. Second, I build on work done by scholars such as Alan Richardson, whose *British Romanticism and the Science of the Mind* (2001) and *The Neural Sublime* (2011) uncover connections between Romantic literature and British theories of mind that emphasize the body. Richardson also introduces cognitive historicism, a new method that uses twentieth- and twenty-first-century cognitive theory to interrogate Romantic concepts.[85]

Models of embodied, embedded, enacted, and extended cognition (the 4 'e's) are particularly apt because, like Romanticism, they emerged in response to Cartesian dualism.[86] Around the beginning of the twenty-first century, cognitive theorists from across the disciplines began to challenge what they saw as hidden dualism in the representational and computational models of cognition that emerged during the cybernetics movement of the 1940s and 50s. Later twentieth-century cognitive scientists and philosophers such as V.S. Ramachandran, Paul and Patricia Churchland, Daniel Dennett, and Ray Jackendoff tended to define cognition as the process of transforming information from the external world into mental representations or symbols that the brain manipulates to internally reproduce knowledge about the world. In this view, the brain processes external inputs in a manner similar to a digital computer in order to produce outputs in the form of language and behaviour. Although these models assume that cognition 'depends ultimately on a whole system—its internal structure and the way it is causally linked to the external world via inputs and outputs', in fact they concentrate cognitive activity in the head, thereby perpetuating a form of mind–body dualism to which 4e

[85] See Gavin Budge, *Romanticism, Medicine and the Natural Supernatural: Transcendent Vision and Bodily Spectres, 1789–1852* (Palgrave Macmillan, 2012); Robert Mitchell, *Experimental Life: Vitalism in Romantic Science and Literature* (Johns Hopkins University Press, 2013); Sharon Ruston, *Creating Romanticism: Case Studies in the Literature, Science and Medicine of the 1790s*, Palgrave Studies in the Enlightenment, Romanticism, and Cultures of Print (Palgrave Macmillan, 2013).

[86] See *The Oxford Handbook of 4E Cognition*, ed. by Albert Newan, Shaun Gallagher, and Leon De Bruin, Oxford Handbooks (Oxford University Press, 2018). For an example of brain-based assumptions about cognition, see the thought experiment in V.S. Ramachandran, 'Mirror Neurons and the Brain in a Vat', *Edge: The Third Culture*, 176 (2006) <https://www.edge.org/conversation/vilayanur_ramachandran-mirror-neurons-and-the-brain-in-the-vat> [accessed 25 November 2008].

theorists object.[87] Proponents of 4e, on the other hand, consider the entire body as 'the system'.

The 4e approach, so named by phenomenologist Shaun Gallagher, is a 'new way of thinking about the mind … inspired by and organised around, not the brain, but some combination of ideas that mental processes are (1) embodied, (2) embedded, (3) enacted, and (4) extended'.[88] Its theorists see cognition as rooted in biological structures beyond just the brain and nervous system, and they hold that cognition can occur only in an embodied subject that is situated in a concrete environment. The four 'e's represent complex, sometimes widely diverse theories about cognition, but most of the debates surround the last two terms, enacted and extended. Where proponents of enactive cognition posit that knowledge and mind emerge from the interaction of the percipient and its surround, extensionists contend that cognitive processes are not confined to the brain or even the body but are distributed across the percipient and its environment.[89] I look to the first three 'e's, embodiment, embeddedness, and enaction, primarily drawing on autopoietic enaction. Originating with Francisco Varela, Evan Thompson, and Eleanor Rosch in the 1990s, enaction posits mutual dependence between percipients and their environments. It theorizes that knowledge production is contingent upon the interaction of the body and the environment, that is, it is enacted. This strain of enactive theory is based on autopoiesis, a biological theory developed by Humberto Maturana and Varela in the 1970s. According to autopoiesis, living organisms are 'defined by … endogenous, self-organizing, and self-controlling dynamics'.[90] They organize themselves from within but not independently of their environment. The boundaries that organisms create constitute and distinguish the 'inner' from the 'outer', thus destabilizing the idea of an absolute subject and object. Unsettling the binary between structuralist and post-structuralist epistemologies, autopoietic enaction defines cognition broadly enough to account for diverse lifeforms. Like the theories advanced by Hartley and Priestley, it destabilizes the strict divisions between species and attempts to unify the multeity into a dynamic system.

[87] Jaegwon Kim, *Philosophy of Mind*, ed. by Norman Daniels and Keith Lehrer, Dimensions of Philosophy (Westview Press, 1996), p. 79.

[88] Mark Rowlands, *The New Science of the Mind: From Extended Mind to Embodied Phenomenology* (Bradford Books, 2010), p. 3.

[89] See Andy Clark and David Chalmers, 'The Extended Mind', *Analysis*, 58.1 (1998), pp. 7–19. Shaun Gallagher, *How the Body Shapes the Mind* (Oxford University Press, 2005) and *Enactivist Interventions: Rethinking the Mind* (Oxford University Press, 2017); Rowlands, *The New Science of the Mind*; Francisco J. Varela, Evan Thompson, and Eleanor Rosch, *The Embodied Mind: Cognitive Science and Human Experience* (MIT Press, 1991).

[90] Thompson, *Mind in Life*, p. 43.

With cognitive historicism as my method and 4e cognition as my theoretical framework, I analyse models of British organicism to revitalize a worn-out critical concept by shifting focus away from the transcendental subject towards the body and the environment in which it is embedded. These paradigms offer a range of theories from dynamic systems theory to neurophenomenology that allows me to make good on Levinson's suggestion that the 'studies of systems and life forms' that originated in the 1960s 'might help us make headway on a question that has been at the heart of critical work in the humanities since the 1980s. Namely, how can the phenomena traditionally assigned to the mind … arise without that transcendental postulate' that plagues eighteenth- and nineteenth-century responses to Cartesianism?[91] Additionally, I use the concept of Romantic structures of engagement to refer to the bodily and environmental structures that enable aesthetic experiences. I define Romantic structures of engagement as an assemblage of structures both internal and external to the body that organize the aesthetic encounters that lead to experiences that are often ineffable and externally unverifiable, such as those described in Romantic poetry.

A diverse set of theories by scholars with varying philosophical and disciplinary investments inspired my concept of Romantic structures of engagement: Raymond Williams, James J. Gibson, and George Lakoff and Mark Johnson. As the phrase suggests, Romantic structures of engagement are akin to Williams's 'structures of feeling', which he formulated to fill a gap in Marxist analytical procedures that treat material culture as if it were a finished product rather than something that is always in process. He describes them as a 'kind of feeling and thinking which is indeed social and material, but … in an embryonic phase'.[92] They refer to potential ideological trajectories that may or may not become hegemonic, and, more importantly, they denote affective, difficult to articulate phenomena that can only be fully understood retrospectively. Similarly, Romantic structures of engagement refer to embodied aesthetic experiences before they are articulated in language and that in the moment 'exert palpable pressures' that influence 'experience and action' and reflect 'specifically affective elements of consciousness' or pre-consciousness.[93] In Romantic-era terms, they refer to the 'spontaneous overflow of powerful feelings' while it is happening.[94]

Gibson, a cognitive psychologist, contributes the idea of environmental affordances. He explains that '*affordances* of the environment are what it

[91] Levinson, *Thinking Through Poetry*, p. 133.

[92] Raymond Williams, *Marxism and Literature* (Oxford University Press, 1977), p. 131.

[93] Williams, *Marxism and Literature*, p. 132.

[94] Wordsworth, 'Preface' to *Lyrical Ballads, Prose Works*, I, p. 127, ll. 129–130.

offers the animal, what it *provides* or *furnishes*' for perception.[95] With respect to Romantic structures of engagement, affordances are structures in the external world that have the potential to engender aesthetic experiences, that is, they are what a given culture collectively defines as an aesthetic object. By leaving the exact objects unspecified, I create space in my aesthetic theory for different historical periods, geographical regions, and cultures that conceptualize aesthetics in various ways, perhaps not even using that term. Thus, my theory encompasses a broad range of objects and experiences that can be considered only within specific contexts. I use the term Romantic structures of engagement to refer to the objects that inspired feelings that the Romantics defined as sublime, beautiful, transcendent, and so forth. However, I also extend Gibson's concept by applying the term affordances to physiological structures in the body that coproduce the aesthetic experiences when they interact with external Romantic structures of engagement. Thus, when I refer to Romantic structures of engagement, I intend to invoke both external and internal affordances.

Lakoff is an anti-Chomskyan cognitive linguist and Johnson is a philosopher who challenges objectivist analytical philosophy. They argue that feelings of transcendence are immanent to the body. In addition to arguing that all mental concepts take their origin in embodied metaphors, they consider transcendence from an embodied perspective, describing it as a physical sense of connection and awe. They also introduce the idea of 'embodied spirituality', which 'requires an aesthetic attitude to the world that is central to self-nurturance, to the nurturance of others, and to the nurturance of the world itself'.[96] In conjunction with their claim that 'artworks provide new ways of structuring our experience' because they 'provide new experiential gestalts and, therefore, new coherences', the concept of embodied transcendence captures my idea that aesthetic experiences occur in the body.[97] The connection they make between spirituality, aesthetics, and nurturance lends credence to my claim that aesthetics can be ethical. Romantic structures of engagement, then, refer to the complex of mind–body–world that I argue makes aesthetic experiences possible.

[95] James J. Gibson, *The Ecological Approach to Visual Perception* (Lawrence Erlbaum Associates, 1986), p. 127.

[96] George Lakoff and Mark Johnson, *Philosophy in the Flesh: The Embodied Mind and Its Challenge to Western Thought* (Perseus Books Group, 1999), p. 566.

[97] George Lakoff and Mark Johnson, *Metaphors We Live By* (University of Chicago Press, 1980), p. 235.

Chapter Summaries

In 'Cognitive Organicism', I situate Hartley's best-known work, *Observations on Man* (1749), in the context of the religious debates occurring among Anglican ministers such as John Gay and Joseph Butler in the wake of Locke's sensationalist psychology. Locke's theory of *tabula rasa* raised questions about people's proclivities towards virtue or vice. Where Butler, close friends with Hutcheson, believed that human beings have an innate moral sense that enables them to know right from wrong, Hartley adopts the position in Gay's *Preliminary Dissertation Concerning the Fundamental Principle of Virtue or Morality* (1731) that each person's conscience develops over time as they accrue experience. Thus, Hartley develops Gay's postulate into a detailed physiological account of cognition that explains, based on the doctrine of association, how people become virtuous or vicious. Although his focus on humankind suggests that Hartley takes an anthropological approach to the study of human history, he, in fact, straddles the line between anthropology and natural history. He challenges Cartesian dualism by envisioning a system in which the mind, body, world, and God are interdependent. I argue that his representation of cognition, which privileges relationality, provides a model of organicism upon which later thinkers drew. I also make the unprecedented argument that Hartley advances an active theory of cognition, which qualifies it as a model of organicism.

'Posthuman Organicism' examines Darwin's intertwined theories of life and mind in *Zoonomia*, *The Botanic Garden* (1789–91), and *The Temple of Nature* (1803). I argue that his historicized account of life, mind, and society, which unites the inanimate and animate world into a system fuelled by social feeling, offers another example of embodied organicism that situates human development firmly in natural history. Drawing on the accounts of sensation and excitability advanced by Edinburgh medical professor William Cullen and his former student John Brown respectively, Darwin hypothesizes an active lifeforce, called sensorial power, that he claims is responsible for sentience as well as life. Sensorial power bridges the Cartesian gap by transmitting sensory data into the body, across the nervous system, to the mind, and, as universal principle, it also blurs the boundaries between humankind and the rest of the living world. I argue that aesthetics is fundamental to Darwin's account of human development, both ontogenetically and phylogenetically. Linking sympathy with ethical action, in *The Temple of Nature* and *The Economy of Vegetation* (1791) Darwin attributes the development of both language and society to the social emotions inherent in the affection parents have for their offspring. However, Darwin does not see society as the special province of human beings but claims that it began with parthenogenic single-celled organisms. Seeing emotion as the basis of subjectivity, he expands it to include all lifeforms. In *The Loves of the Plants* (1789), he anthropomorphizes plants to demonstrate that they not only experience emotions, such as love, but they are subject to the whole range of cognitive processes. Using Thompson's account of autopoietic enaction in *Mind in Life* (2007) as my theoretical frame, I argue that Darwin advances a model of embodied organicism that vitalizes nature and decentres human beings in ways that offer a potent posthuman paradigm for conceptualizing humanity's place in the cosmos.

In 'Phenomenological Organicism', I argue that Coleridge's model of life and mind, advanced in 'Theory of Life' (composed 1816) and *Biographia Literaria* (1817), relies more heavily on the body than critics have recognized. Drawing on Maurice Merleau-Ponty's work, I maintain that Coleridge invites us to read 'Theory of Life' phenomenologically rather than ontologically, thus challenging scholars who cast Coleridge as a dualist despite his aversion to Cartesianism. My discussion of passages from *Biographia* in conjunction with notebook entries establishes the importance of the body to his concept of organicism. Read phenomenologically, Coleridge's account of life connects the physical, vital, and human orders—to use Merleau-Ponty's terminology—to synthesize his sense that the world is simultaneously multiple and unified. Drawing on dynamic systems theory, I also provide a new way to understand Coleridge's attempt to preserve freewill. Drawing on Romantic structures of engagement, I contend that these texts imply that aesthetic experiences are embodied. Turning to his essays on method in *The Friend* (1818), I contend that the embodied and embedded mind that Coleridge implicitly posits creates the necessary conditions for discovering the truths that he claims can only be accessed from within. Aesthetics, for Coleridge, is not just about form but about active engagement with the external world for the purpose of exercising his method, which I demonstrate with my analysis of 'The Eolian Harp'.

In 'Ethical Organicism', I analyse *The Excursion* (1814), *The Prelude* (1805), and a selection of prose, including the Prefaces to *Lyrical Ballads* (1800, 1802) and *Poems* (1815) to elucidate the relationship between Wordsworth's theories of cognition and organicism. I situate his thinking in the context of his reading of British and French materialist philosophers to argue that he adapted their ideas to formulate his concept of embodied and embedded cognition, which serves as a model for his organicism. My analysis of the imagery and structure of key passages in *The Prelude* and *The Excursion* expounds Wordsworth's theory of cognition. I contend that he was interested in Cartesianism as a metaphor that structures people's worldviews and, in consequence, their attitudes and behaviours. He offers organicism as an alternative way to experience and envision the world in terms of relationality rather than separation. His poetry and prose call attention to the compatibility between bodies and nature, which work together to produce thoughts, feelings, and attitudes that have potential to alter behaviour. Cognition is not a principle for Wordsworth but an active process, and organicism is a way to conceptualize how self-other relationships and interactions are structured. Wordsworth's ethics, I contend, were heavily influenced by Holbach's view of nature and its relationship to the moral order. Like Adam Potkay, who asserts that Wordsworth's ethics were rooted in encounters with others, I argue that the Wanderer in *The Excursion* exemplifies his ethical organicism and provides readers with a model to emulate by illustrating a politics of acting locally. I argue that Wordsworth's poetry provides a map, model, and motivation to inhabit the body and experience it as embedded in a web of relations by having deeply embodied encounters that elicit awe, wonder, synchrony, among other aesthetic experiences.

My analysis of key texts by Hartley, Darwin, Coleridge, and Wordsworth, which I treat as case studies, not only produces new insights into organicism as a critical and aesthetic concept but also speaks productively to our current moment. Like Steigerwald's investigation into the experimental roots of German vitalism, this study has been 'prompted', in part, 'by changing scientific discourses and their epistemic commitments'.[98] However, this is not a study in presentism. I do not claim that Romantic theories of cognition prefigure, anticipate, or reflect our current moment. Rather, following Miranda Anderson's lead in *The Renaissance Extended Mind* (2015), I argue that debates and theories of mind in any historical period, including our own, 'reflect enduring or changing aspects of the paradigm[s]' by which we understand cognition.[99] In contrast to the narrative that knowledge is progressive, I contend that the ways that a society understands the mind, cognition, and how they relate to the body and the world come and go. Paradigms can and do reappear in altered and revised forms at different points in history. Paradigmatic similarities between the past and present are an invitation to interrogate the differences between the periods and to examine the conditions of production that engendered them. According to Steigerwald, 'rethinking the past also does critical work, in prompting a rethinking of the present in relationship to its past and to its possible futures'.[100] Even though those of us living two centuries after the Romantics encounter challenges far different from those faced by citizens living under Britain's repressive governmental regime after the failure of the French Revolution—late-stage global capitalism, climate change, environmental racism, the revival of fascism—I maintain that the insights offered by embodied organicism help us to see the significance of certain twenty-first-century trends more clearly and that they empower us to find efficacious responses to our own conditions of modernity. Furthermore, I claim that the problems engendered by Cartesian dualism to which Romantic thinkers responded continue to plague us today, particularly the separatist

[98] Steigerwald, *Experimenting at the Boundaries of Life*, p. 25.

[99] Miranda Anderson, *The Renaissance Extended Mind* (Palgrave Macmillan, 2015), p. vii.

[100] Steigerwald, *Experimenting at the Boundaries of Life*, p. 35.

worldview it metaphorically models. I take up these issues in the conclusion and address the implications and relevance of Romantic organicism in the twenty-first century. As I interrogate the similarities and differences between our historical moment and the one that I write about in these pages, I stake a claim for Romantic aesthetic categories but refashioned for the twenty-first century. Despite, the materialist turn in science in the mid-twentieth century and the turn to cultural materialism in the 1980s, I contend that we continue to be dominated by the Cartesian metaphor that encourages us to see ourselves as isolated, individual subjects who have no real connection to the world in which we live and which we are ostensibly just passing through. I conclude with a discussion of Jerome Christensen's concept of Romantic hope. As Levinson notes, this question of parts and wholes does not have to be conceived of metaphysically, and my goal is to prove her right.

A Note on Bodies and Embodied Cognition

Studies such as this one that talk abstractly about 'bodies' run the risk of replicating the errors of Enlightenment universalism: when its advocates proclaimed that all men have natural, inalienable rights, they literally meant men, and not even most of them. This risk is exacerbated by the fact the cognitive science retains a similar bias. Most of its examples seem to be unconsciously gendered. For example, Ramachandran's brain-in-a-vat thought experiment (see footnote 86, above) overlooks sex, race, and gender, and the terms of the experiment feature mostly White men. Even 4e cognitive theories appear to suffer from this bias. Consider Andy Clark and David Chalmers's illustration of extended cognition, which features a man named Otto. However, Ralph James Savarese argues that theories of situated cognition create space for considering cognition for an expanded range of bodies. He mentions people with 'congenital deafness or autism or even impaired mobility', arguing that different types of bodies 'produce ... different form[s] of cognition'.[101]

This is to say that I am aware that in talking about bodies in the abstract, in theories by British White men no less, that this project is in danger of repeating old mistakes. After much retraining, I now have no 'normative' body or subject in mind when I hear, think, or talk about the body—at least not most of the time. I invite you to envision the abstract body in any way that makes sense to you. What is more, I invite you to become consciously aware of the body that comes to mind as you read my words. My goal, as a person who would likely have been excluded from the theories and aesthetic experiences I discuss

[101] Ralph James Savarese, 'Cognition', in *Keywords for Disability Studies*, ed. by Rachel Adams, Benjamin Reiss, and David Serlin (New York University Press, 2015), pp. 40–42 (pp. 40, 42).

or who would have maybe even served as an experimental subject, is to make them more widely accessible. This is one of the main reasons that I find 4e theories of cognition appealing, despite the blind spots of some of its theorists.

The 4e approach to cognition foregrounds the body, thus validating forms of knowledge that fall outside of the narrow boundaries of ratiocination and broadening the definition of cognition to include emotion, desire, and other types of embodied phenomena that cognitive scientists and behaviourists in the 1950s and 1960s black-boxed. It also challenges cognitive chauvinism and decentres the human in productively posthuman ways, much as, I argue, Darwin does in his work. Nonetheless, critics such as Richard Sha and Joel Faflak, in their introduction to *Romantic Consciousness, Revisited* (2022), raise valid concerns. They argue that 4e theories endorse an ideology of efficiency, foreclose Romantic alienation and disjuncture, and disable scepticism, all of which prioritize thinking-as-doing over critical thought.[102] However, embodied, embedded, and even enacted cognition do not automatically preclude abstract thought. Furthermore, models of embodied cognition are metaphorically necessary alternatives to representational and computational models.

In contrast to the thought-as-action paradigm, which focuses on patterns that link sensory stimulation and movement, biological enaction attends more to sense-making.[103] The emphasis is on the relationship between sensation and cognition, which is made possible and structured by bodily movement. Hartley's primitive argument that pleasure and pain attach to all sensation, which gives it a *de facto* emotional valence (pleasure/happiness, pain/unhappiness), underscores the fact that no adequate account of cognition can bypass the body. Additionally, accounts of embodied cognition do not deny the crucial role of the central and peripheral nervous systems, rather they draw attention to the body beyond the brain and nerves and to the nuanced ways it receives, evaluates, and processes perceptual data. In short, they emphasize the role of the body and how it affects our thinking, including abstract reasoning, which has too long been ignored.

The long Cartesian tradition of vilifying the body still justifies associating it with marginalized populations who are believed to feel—both emotionally and physically—less, be more capable of manual labour, and are, consequently, subject to socially sanctioned abuse. Beginning with Plato, Western philosophy has feminized the body, which has justified centuries of sexism. Simone de Beauvoir confronts this dichotomy of the masculinized mind and the feminized

[102] Richard C. Sha and Joel Faflak, 'Introduction: *Romanticism and Consciousness* Redux', in *Romanticism and Consciousness, Revisited*, ed. by Richard C. Sha and Joel Faflak (Edinburgh University Press, 2022), pp. 1–14 (pp. 8–11).

[103] For the difference between biological and sensorimotor enaction, see Dave Ward, 'Enactivism', *A History of Distributed Cognition*, November 2024 (University of Edinburgh) <http://www.hdc.ed.ac.uk/seminars/enactivism> [accessed 7 February 2015].

body in *The Second Sex* (1949) and 'points out the bad faith' obfuscation that underwrites the undetermined and free Cartesian subject.[104] In Beauvoir's analysis, according to Judith Butler's perceptive interpretation, only man can be a free thinking subject because his 'disembodiment is only possible on the condition that women occupy their bodies as their essential and enslaving identities', which bars women 'from identifying with '"consciousness", that transcending activity unrestrained by the body'.[105] In other words, some men can play at Cartesian subjectivity because women, and I would add people of colour, bear the burden of embodiment. The myth of the detached *cogito* can be a lived reality only for those who do not have to think about or deal with the quotidian needs of the body. While Sha and Faflak are right to worry that theories of embodied cognition privilege uncritical action over critical thought, there is nothing inherent in the processes of embodied cognition that prevent abstract thought. However, ignoring the body's role in cognition privileges the ideology outlined above.

Additionally, embodied cognition allows for the possibility of pre-discursive knowledge. Despite the important insights generated by post-structuralist theory, the claim that all knowledge is discursive, I contend, is both false and dangerous. In *Focusing* (1982), Eugene Gendlin discusses a form of knowing that he calls 'felt sense', which he describes as 'bodily awareness of a situation, person, or event ... that encompasses everything you feel and know about the given subject at a given time' that gets communicated 'all at once rather than detail by detail'.[106] Distinguishing felt sense from thoughts and emotions, he explains that 'it has emotional components in it, along with factual components', but it is 'more complex' than a 'single emotion'.[107] While it is possible to verbalize felt sense, even without articulation it provides valuable information. For example, when girls express a sense of unease that they cannot put into words about a 'creepy' person or situation, they are often discounted, which over time teaches them to ignore their non-discursive embodied knowledge and can lead them into dangerous situations.[108] Peter A. Levine, who holds doctorates in psychology and in medical biophysics, contrasts felt sense with categorical emotions such as anger or happiness,

[104] Judith Butler, 'Sex and Gender in Simone de Beauvoir's *Second Sex*', *Yale French Studies*, 72 (1986), pp. 35–49 (p. 43).

[105] Butler, 'Sex and Gender', p. 43.

[106] Eugene T. Gendlin, *Focusing* (Random House Publishing Group, 1982), p. 37.

[107] Gendlin, *Focusing*, pp. 37–38.

[108] See Gavin de Becker, *The Gift of Fear: Survival Signals That Protect Us from Violence* (Dell, 1998) and *Protecting the Gift: Keeping Children and Teenagers Safe*, reprint edn (Dell, 2000); Harris O'Malley, 'Men Really Need to Stop Calling Women Crazy', *Washington Post*, 9 July 2014 <https://www.washingtonpost.com/posteverything/wp/2014/07/09/men-really-need-to-stop-calling-women-crazy/> [accessed 30 June 2024].

explaining that it is 'based on physical sensations and encompasses the whole inner landscape of our bodies'.[109] It also refers to the 'contours of feeling ... that direct the movement of our life throughout the day'.[110] Thus, theories of embodied cognition make space for important and potentially life-saving non-discursive forms of knowing.

Finally, embeddedness and enaction speak to the limitations of human knowledge, or rather they imply that all knowledge is phenomenological, which as Sha notes, challenges species chauvinism. In *Metaphors We Live By* (1980), Lakoff and Johnson position their theories of embodied language against objectivist theories of knowledge, whose proponents believe that human beings have access to absolute and objective knowledge of the external world. Similarly, Varela, Thompson, and Rosch challenge the notion that knowledge exists transcendentally or that it is 'out there' in the world waiting to be discovered, arguing instead that it is contingent on physiology. Accounts of embodied and embedded cognition acknowledge that there is a so-called objective world with which we interact but holds that we do not have unconditional or unconditioned access to it. We are not Ralph Waldo Emerson's 'transparent eyeball' but organisms with eyeballs structured in particular ways that shape how and what we see.[111] Levinson explains the action-perception feedback loop in a way that makes it clear that it does not necessarily bypass critical thinking, stating that if 'action is inseparable from perception' then 'perceiving means perceiving as; perceiving in reference to a world'.[112] Thus, embeddedness and enaction speak to how human beings come to understand a world that discloses itself to them through their specific sensory and cognitive apparatus in their specific environment.

Sha contends that 'if enactivism heals the Cartesian separation of mind and body and mind and world insofar as both object and subject are replaced as a verb, this healing comes, at least in theory, at the considerable costs of denying the value of abstraction (and distance), and irony, along with the value of the subjective feel of experience', a point with which I agree.[113] It is

[109] Peter A. Levine, *What Is Felt Sense?*, dir. by Somatic Experiencing International, YouTube, 2019, loc. 4:11–18 <https://www.youtube.com/watch?v=ZBLe84U7AaM> [accessed 28 June 2024]. Levinson also briefly discusses Gendlin in *Thinking Through Poetry*, see n. 62, p. 23 and n. 50, p. 251.

[110] Levine, *What Is Felt Sense?*, loc. 3:35–50.

[111] Ralph Waldo Emerson, 'Nature', in *Nature and Selected Essays* (Penguin, 2003), pp. 35–84 (p. 39). The canonical example in enactive cognition is the way different species see colour, which depends on the number of colour processing channels in the eye (tetrachromatic, trichromatic, dichromatic, for example).

[112] Levinson, *Thinking Through Poetry*, p. 134.

[113] Richard C. Sha, 'Blakean Experience and the Hard Problem of Consciousness Revisited', in *Romanticism and Consciousness, Revisited*, ed. by Richard C. Sha and Joel Faflak (Edinburgh University Press, 2022), pp. 94–117 (p. 97).

imperative that we ensure that mental phenomena such as these do not get theoretically divorced from cognition. Rather than replacing abstract thinking, I argue that 4e cognition provides a necessary corrective to nearly four hundred years of Cartesian disconnection, alienation, chauvinism (and not just against other species but also other cultures, genders, and so on), and their material effects. As the following pages reveal, I am more interested in the metaphors that these paradigms of cognition offer and their potential to consciously or unconsciously affect the way we understand, experience, and behave in the world. These theories stand as a reminder that the 'self is never not enmeshed in—indeed, engendered ... by—collectivity'.[114]

[114] Levinson, *Thinking Through Poetry*, p. 18.

Chapter One

Cognitive Organicism

David Hartley

Even as a child, David Hartley was interested in how the various parts of the world function together to form a cohesive unity. When he was a boy of just nine years old, he used to swing 'backwards and forwards upon a gate … meditating on the nature of his own mind; wishing to find out how man was made; to what purpose, and for what future end'.[1] His daughter's apocryphal account of his childhood musings provides a glimpse into the mutually dependent theories of cognition, theology, and moral progress that he developed in adulthood. Wondering about how the mind is connected to the body ('how man was made'), how physiology shapes humankind's *raison d'etre* ('to what purpose'), and how it all connects to the biblically promised afterlife ('for what future end'), Hartley poses questions about 'Connexions and Influences', that is, relationships. He wanted to understand not just how the world works but how its various parts are connected and, as the son of an Anglican minister, how they relate to religious truths. He attempts to answer these questions in his account of cognition detailed first in *Conjecturae Queadam de Sensu, Motu, et Idearum Generatione*, which he appended to his 1746 treatise on kidney stones.[2] A few years later, he responded more fully to

[1] Rebecca Warner, *Original Letters, from Richard Baxter, Matthew Prior, Lord Bolingbroke, Alexander Pope, Dr. Cheyne, Dr. Hartley, Dr. Samuel Johnson, Mrs. Montague, Rev. William Gilpin, Rev. John Newton, George Lord Lyttleton, Rev. Claudius Buchanan, Etc. Etc.* (Richard Cruttwell, 1817), p. 93 <https://www.google.com/books/edition/Original_Letters_from_Richard_Baxter_Mal/-RKZT6wUA98C?hl=en&gbpv=0> [accessed 24 May 2010].

[2] Hartley's treatise was not translated into English until 1959. See David Hartley, *Various Conjectures on the Perception, Motion, and Generation of Ideas*, trans. by Robert E.A. Palmer (University of California Press, 1959).

these questions in *Observations on Man, His Frame, His Duty, and His Expectations* (1749), which builds on John Locke's sensationalist psychology in *An Essay Concerning Human Understanding* (1690, 1694, 1695, 1700, 1706) and indirectly challenges the theological and scientific implications of Cartesian dualism.[3]

Bifurcating the world into immaterial and material substances and associating God and the human mind with the former and everything else in the phenomenal world with the latter, René Descartes substantially divorces God (spirit) from his creation (matter) and raises questions about how the mind interacts with the body and the world.[4] Many religious thinkers welcomed Cartesian dualism despite this obstacle because it easily accounts for the immortality of the soul. However, when Locke challenged Descartes's epistemology with the claim that the '*idea of God be not innate*' and, therefore, '*no other* [*idea*] *can be supposed to be innate*', it raised the spectre of materialism despite Locke's explicit assurances that he was a dualist.[5] According to Peter H. Sedgwick, in the wake of Locke's *Essay* Anglican theological debates moved away from topics such as ecclesiology and cosmology towards questions of morality and moral sense.[6] Two camps formed in response to sensationalist psychology. One included Joseph Butler and Thomas Reid, who adopted the '"immaterial substance" theory of self against Locke'.[7] They agreed with Francis Hutcheson's claim in *An Essay on the Nature and Conduct of the Passions, with Illustrations on the Moral Sense* (1728) that human beings have

[3] *Human Understanding* went through four editions during Locke's lifetime and one shortly after his death. Each one contained substantive changes and additions, including the discussion of association which did not appear until the fourth edition of 1700.

[4] Despite Descartes's attempt to answer queries about interactionism from Princess Elisabeth of Bohemia and others by claiming that the soul is connected to the body via the pineal gland, he did not convince his critics. In *Treatise on Man* (published posthumously 1662, 1664) and in *The Passions of the Soul* (1649), he asserts that the 'activity of the soul consists entirely in the fact that simply by willing something it brings it about that the little gland to which it is closely joined moves in the manner required to produce the effect corresponding to this volition', but this explanation does not account for how two discrete substances interact. René Descartes, *The Philosophical Writings of Descartes*, trans. by John Cottingham, Robert Stoothoff, and Dugald Murdoch, 2 vols (Cambridge University Press, 1984), I, p. 343. See Gert-Jan Lokhorst, 'Descartes and the Pineal Gland', *The Stanford Encyclopedia of Philosophy*, Winter 2021 (Metaphysics Research Lab, Stanford University), <https://plato.stanford.edu/archives/win2021/entries/pineal-gland/> [accessed 14 March 2025].

[5] John Locke, *An Essay Concerning Human Understanding*, 4th edn (Prometheus Books, 1995), p. 50.

[6] Peter H. Sedgwick, *The Development of Anglican Moral Theology, 1680–1950* (Brill, 2024), p. 131.

[7] Bob Tennant, *Conscience, Consciousness, and Ethics in Joseph Butler's Philosophy and Ministry* (Boydell & Brewer, 2011), p. 12.

an innate sense of right and wrong, while the other group, inspired by John Gay, 'became deeply affected by the influence of Locke's epistemology and ethics' and held that 'conscience was only acquired as human beings grew in maturity', a theory that was accused of being materialist.[8]

Despite his friendship with Butler and his unwillingness to commit to materialism, Hartley adopted Gay's theory of moral sense as outlined in his *Preliminary Dissertation Concerning the Fundamental Principle of Virtue or Morality* (1731), which was prefixed to Edmund Law's translation of William King's 1702 *An Essay on the Origin of Evil.* Hartley draws on these two works as well as on Isaac Newton's *Mathematical Principles of Natural Philosophy* (1687) and *Opticks* (1704). In accordance with his daughter's account of him as a child, in *Observations* Hartley addresses the theological issues that concerned him. By this point, he had rejected the dualist implication that substantial differences separated God from his creation, but he also believed that humankind's first disobedience had alienated humans from God. Indeed, Hartley had refused to become an Anglican minister because he did not agree with all thirty-nine articles of faith and could not accept the doctrine of eternal damnation, which holds that God will punish sinners by damning them to hell, a place of torment and separation. These three apprehensions motivated him to use the problem of mind–body interaction, Locke's psychology, Gay's account of conscience, King's account of evil, and Newton's brief discussion of vibrations, along with observations from his medical practice to think through the relationship between God, humankind, and nature.

In contrast to Thomas McFarland, who describes Hartley's work as 'a compendium of Locke and Newton that attempted, with heroic inconsequence, to cover the philosophy of "it is" with a fretwork of pious Christian avowals', I argue with R. C. and Lady Kathleen Oldfield, Barbara Bowen Oberg, and C.U.M. Smith that his synthesis of others' ideas resulted in an original theory that deserves reconsideration.[9] I contend that the scholars who have dismissed his theory as mechanistic and passive have not looked carefully enough at his system. I argue that Hartley advances a theory of active cognition that belongs to a tradition of British organicism. Participating in the natural theology movement that held that humans could gain knowledge of God by studying nature and exercising reason, he hypothesizes that mind, body, and nature form an interdependent system that connects human beings to their creator. He bases his model on the body, the parts of which function together to enable cognition, which makes it possible for people to be aware of and commune with God. Hartley's detailed account of cognition illustrates his cosmology,

[8] Sedgwick, *The Development of Anglican Moral Theology*, pp. 128, 132.

[9] Thomas McFarland, *Coleridge and the Pantheist Tradition* (Clarendon Press, 1969), p. 169. See also Barr; Fairchild; Hayden; and Allen, n. 4, p. 424.

which envisions the universe as an interactive, mutually dependent system comprising God, the phenomenal world, and the people in it.

In both volumes of *Observations*, Hartley elaborates an account of human development that starts with an embodied and embedded mind that enables people to enact knowledge of God. Drawn from twenty-first-century cognitive theory, these concepts can be used to elucidate his non-dualistic model of cognition, illustrating how it is active and how knowledge of God can emerge from the interaction of percipients and their environment. Considering nature and the human body as expressions of God's truth, Hartley uses his 'scientific inquiries' not only to provide a 'comprehensible background to revealed religion' but to investigate Christian teleology.[10] His attempt to think the phenomenal world as a holistic system is also an attempt to heal the schism precipitated by the fall, which he saw not as ontological but as epistemological. For this reason, cognition plays a crucial role in Hartley's understanding of humankind's relationship to God, before the fall, after the second coming, and in the postlapsarian world. To account for how experience leads individuals to become vicious or virtuous, Hartley contends that character formation occurs through embodied cognitive processes. His account of moral development draws on King's argument about the relationship between virtue and vice. The relationship that he theorizes between the mind, body, nature, and God implies a system of relationality that exemplifies organicism. In the first part of this chapter, I flesh out Hartley's theory, examining how it reveals a commitment to relationality over ontology. By focusing on aspects of Hartley's theory other than its latent materialism, I uncover more important implications, such as the body's active role in constructing knowledge, not just of the world but also of God.

In the second part of the chapter, I discuss Hartley's moral philosophy, in which he applies Locke's ideas about association to Gay's discussion of virtue. His explanation of moral development, the result of cognitive association, provides insight into Romantic era aesthetics, particularly how such experiences can be conceived as ethical, even though he disagrees with Hutcheson's claim that humankind has an inner sense that recognizes beauty. Aesthetics and the imagination, its corollary in the human mind, are fundamental to Hartley's theory in which the latter serves as the cornerstone of moral maturity. As people pass through the six intellectual affections, or stages of mental and moral development, they grow closer and closer to God until they reunite with him, to the extent possible in a fallen world. Hartley speculates that if humanity as a whole could successfully direct association, the majority of the population would reach the highest stage of development, moral sense, or disinterested love

[10] R.C. Oldfield and Lady Kathleen Oldfield, 'Hartley's *Observations on Man*', *Annals of Science*, 7.4 (1951), pp. 371–382 (p. 373).

of God and humankind. Then the promise of Hartley's cognitive organicism would be fulfilled and unity would be restored to God's creation.

Although he works within a Newtonian framework, Hartley's effort to understand the 'mutual Connexions and Influences' of the mind and body in response to accounts of cognition that pit the two against each other produce surprisingly nuanced solutions to the Cartesian dilemma, despite his commitment to necessity. Hartley seems to take an anthropological approach to understanding how human beings came to be in their current state and how they can repair it, yet his emphasis on humankind's embeddedness in the natural world places him within the tradition of natural history. He serves as an example of a thinker who privileges human cognition, including the capacity for self-awareness, without separating humankind from nature or from other species. His attention to the relationship between the mind, body, and the world establishes an infrastructure of relationality that challenges the Cartesian separation of spirit and matter. Additionally, Hartley resituates God in the context of creation, positing that nature connects him to his creatures. His emphasis on the importance of aesthetics to morality, not through an innate moral sense but through developmental processes, informed and helped define Romantic ways of seeing and making sense of the world. Thus, Hartley serves as an important model of British embodied organicism.

Relationality and Epistemological Fractures

Hartley and nearly every other philosopher who preceded him, from Plato onwards, believed that in the beginning humankind lived in a state of perfect freedom and unmediated contact with the pure good from which the universe originated—until something happened.[11] For Christians that something was original sin, when the first humans fell from grace by directly disobeying God's injunction not to eat fruit from the tree of knowledge of good and evil. For Plato and his followers, the fall was a descent into matter, and they viewed the material world, if not as outright evil, as a distraction from the good to which they aspired to return. After Descartes introduced substance dualism, Christians tended to share this view of matter. They regarded the body as subject to the temptations that abounded in the phenomenal world. In *Disquisitions on Matter and Spirit* (1777), Joseph Priestley claims that Cartesianism strengthened the 'philosophical notion of matter being the source

[11] According to Jacques Derrida, 'all metaphysicians, from Plato to Rousseau, Descartes to Husserl, have proceeded in this way, conceiving good to be before evil, the positive before the negative, the pure before the impure, the simple before the complex, the essential before the accidental, the imitated before the imitation, etc.' Jacques Derrida, *Limited Inc*, ed. by Gerald Graff, trans. by Jeffrey Mehlman and Samuel Weber (Northwestern University Press, 1988), p. 236.

of evil', charging Descartes with being the first person to claim that God is connected to the world 'only by his *knowledge* and *power*, having *no relation to place*'.[12] By positing God as a radically different substance from human beings, excepting the mind, Cartesianism separates him wholly from his creation. Thus, Cartesian metaphysics parallels the fall, which severed the union between God and humankind, shattered prelapsarian wholeness, and divorced the material world from the divine. In this way, it reinforces the idea that the body and its environment are inherently sinful.

Where most post-Cartesian Christian thinkers, such as the Cambridge Platonists, understood the fall as an ontological break, Hartley did not.[13] Instead, he saw it as an epistemological fracture that disrupted humankind's uninterrupted communion with God and perfect comprehension of his truth. In Hartley's view, prior to the original sin, nature 'functioned to lead man directly from knowledge of created nature to God'.[14] Thus, the fall did not fundamentally change the physical world, the human body, or the mind, which enabled people to know God. Rather, Adam and Eve metaphorically shattered the glass through which humans perceived the world, which resulted in a distorted apprehension of God and his system. When 'mankind acquired knowledge of good and evil, he lost his innocent, integrated vision of the world'.[15] Unfettered knowledge of God was lost in error, which prevents humans from perceiving and talking about him clearly. From that time, according to Hartley, people were 'utterly unable to think or speak properly of him' but could only talk about God in metaphorical and misleading ways.[16] For example, depictions of him as a loving father or divine judge 'fall infinitely short of the Truth, of representing the Deity as he is, [and] are mere popular and anthropomorphitical Expressions'.[17] Furthermore, the dualist claim that God bears no spatial relation to the phenomenal world is 'an inconsistent

[12] Joseph Priestley, *Disquisitions Relating to Matter and Spirit* (Kessinger Publishing, 2005), pp. 345, p. 191.

[13] Ralph Cudworth, a leading Cambridge Platonist, 'combines atomist natural philosophy with Platonic metaphysics', which he identified 'as having originated with Moses', who transmitted it to the Greeks, 'including Descartes whom Cudworth regarded as a reviver of Mosaic atomism'. Sarah Hutton, 'The Cambridge Platonists', in *The Stanford Encyclopedia of Philosophy*, ed. by Edward N. Zalta and Uri Nodelman, Summer 2024 (Metaphysics Research Lab, Stanford University) <https://plato.stanford.edu/archives/sum2024/entries/cambridge-platonists/> [accessed 21 May 2024].

[14] Robert Marsh, 'The Second Part of Hartley's System', *Journal of the History of Ideas*, 20.2 (1959), pp. 264–273 (p. 273).

[15] Margaret Leslie, 'Mysticism Misunderstood: David Hartley and the Idea of Progress', *Journal of the History of Ideas*, 33.4 (1972), pp. 625–632 (p. 623).

[16] David Hartley, *Observations on Man, His Frame, His Duty, and His Expectations*, 2 vols (S. Richardson, 1749), II, p. 8.

[17] Hartley, *Observations*, II, p. 42.

Fiction' based on a 'feign[ed] ... Resemblance between material and immaterial Beings' because humans 'cannot ... discover any Relation which Space or Place bear to Divine Existence'.[18] Similarly, Hartley considered the dualist idea that humans can know God with their immaterial minds to be epistemologically impossible for fallen beings.

Despite Hartley's oblique criticism of dualism, the materialist implications of his theory made him uncomfortable. According to Oberg, he 'conscientiously wished to avoid materialism', probably because he knew it would be followed by accusations of atheism.[19] Even writers who explicitly espoused and wrote defences of Christianity, such as Priestley, were still labelled atheists by their adversaries. While Hartley's account of cognition in the first part of *Observations* seems undeniably materialist, his discussions of ontology are nonetheless ambiguous. Generally, he avoids directly discussing it and his occasional mentions tend to be brief and avoid committing either to materialism or dualism. For instance, towards the end of the first volume, he addresses Locke's suggestion that God could make matter think if he so desired. Hartley makes it clear that he 'no-ways presume[s] to determine whether Matter can be endued with Sensation or no' because it 'is a Point foreign to the Purpose of [his] Inquiries'.[20] This comment indicates that he has no desire to enter into debates about panpsychism, in part because it is irrelevant to his theory of mind. Hartley also claims that ontology does not resolve the fundamental question about whether matter is inert or capable of self-motivated activity. Descartes and his followers argued the former, insisting that it cannot move itself. Thus, the immaterial substance is the source of movement in the phenomenal world. Staunch materialists such as the radical French philosopher Baron d'Holbach claimed that nothing but 'matter and motion' comprise the universe, and that both are material because motion inheres in matter.[21] Hartley addresses the problematical relationship of matter and motive by stating that he 'see[s] clearly, and acknowledges readily, that Matter and Motion, however subtly divided, or reasoned upon, yield nothing more than Matter and Motion still'.[22] In other words, regardless of how they are defined ontologically, the two fundamental elements remain as they are and therefore make no difference to him.

Hartley's most direct statement against dualism appears in a response to Butler's assertion that 'Consciousness is a single and indivisible Power', which

[18] Hartley, *Observations*, II, p. 34.

[19] Barbara Bowen Oberg, 'David Hartley and the Association of Ideas', *Journal of the History of Ideas*, 37.3 (1976), pp. 441–454 (p. 443).

[20] Hartley, *Observations*, I, p. 512. See John W. Yolton, *Thinking Matter: Materialism in Eighteenth-Century Britain* (University of Minnesota Press, 1983), pp. 14–28.

[21] Holbach, *System*, I, p. 30.

[22] Hartley, *Observations*, I, p. 512.

is a core feature of the liberal humanist subjectivity that Cartesianism underwrites.[23] Hartley claims that 'the Unity of Consciousness seems to [him] an inconclusive argument', and yet even that statement is ambivalent, refusing to directly deny its plausibility.[24] Scholars such as R.B. Hatch suspect that Hartley 'had his own doubts about the immaterial character of the mind' even though he seemed to be 'opposed to discarding the concept'.[25] Worried that he would be perceived as denying the soul's immortality, he tries 'to avoid the dangerous implications of his doctrines by ignoring them'.[26] Even in the second volume, where Hartley turns to religion and virtue formation, he undermines his explicit claim that one of God's attributes is immateriality by adding that it is not his most important quality, nor does it matter whether Christians hold this belief. More important than immateriality, according to Hartley, is God's omnipotence and benevolence. He asserts that 'it ought to be observed, that if a Person acknowledges the infinite Power, Knowledge, and Goodness of God, the Proofs of which are prior to, and quite independent on [*sic*], that of his Immateriality, this Person acknowledges all that is of practical Importance'.[27] It is clear that Hartley does not divide the universe into material and immaterial realms. As Richard C. Allen notes, he 'makes no claim that they refer to two different orders of reality, the one material and the other spiritual', but rather he sees human beings and the world as 'both material *and* spiritual'.[28] Additionally, 'nowhere in *Observations* does Hartley make use of the concept of an immaterial substance as an explanatory principle'.[29] Despite the critical commonplace that Hartley is a materialist, I maintain his ambivalence towards ontology demonstrates that it has no impact on his theory. The relationships he sets up between the mind, body, and world are valid under any system. Furthermore, focusing on other aspects of his theory reveals the more important implication that it is undergirded by profound relationality.

In contrast to Cartesian accounts of cognition, Hartley's explanation rests on the premise that God created the world with one goal—blissful communion with him. To this end, nature and the body facilitate his divine will for his creation

[23] Joseph Butler, *The Analogy of Religion, Natural and Revealed, to the Constitution and Course of Nature* (James, John and Paul Knapton, 1736), p. 16, Eighteenth Century Collections Online <https://link.gale.com/apps/doc/CW0118843279/ECCO?sid=bookmark-ECCO&xid=bd928775&pg=5> [accessed 22 April 2024].

[24] Hartley, *Observations*, I, p. 512.

[25] R.B. Hatch, 'David Hartley: Freewill and Mystical Associations', *Mosaic*, 7.4 (1974), pp. 29–39 (p. 31).

[26] Oberg, 'David Hartley', p. 444.

[27] Hartley, *Observations*, II, p. 33.

[28] Richard C. Allen, *David Hartley on Human Nature* (State University of New York, 1999), p. 404.

[29] Allen, *David Hartley*, p. 181.

to be happy. The claim that God wants people to be happy, which Hartley encountered in Gay's *Dissertation*, was 'theologically controversial for his time', which was dominated by dour Calvinist and Presbyterian worldviews.[30] Gay asserts 'that it is evident from the Nature of God, *viz.* of his being infinitely happy in himself from all Eternity, and from his Goodness manifested in his Works, that he could have no other Design in creating Mankind than *their* happiness'.[31] Although Gay controverts Hutcheson's notion of innate moral sense, they agree that virtuous action consists of promoting happiness, a tenet that became widespread in moral philosophy and in natural religion. Proponents of natural religion, such as Priestley, also a man of science, saw the compatibility between humankind and nature as confirmation of God's will.

Priestley teaches that 'it is a considerable evidence of the goodness of God, that the inanimate parts of nature, as the surface of the earth, the air, water, salts, minerals, &c. are adapted to answer the purposes of vegetable and animal life' by 'contributing to the support of animal life'.[32] He reasons that God must want his creatures to be happy since it makes no sense that a loving God would delight in their misery. Gay also appeals to reason, contending that the 'Criterion of Virtue' dictates that 'acting agreeably to *Nature, or Reason*, (when rightly understood) would perfectly coincide with the *Fitness of things*; the Fitness of things ... with *Truth*; Truth with the *Common Good*; and the Common Good with the *Will of God*'.[33] Hartley also held that God created the natural world and human beings to work together to realize his will. Far from being a temptation that leads people away from God, the body registers the effects of virtue and vice, which encourage people to pursue the less painful path. He argues that God endowed humankind with 'such Faculties, as, by their natural Workings, make them take Pleasure in this Contemplation of his infinite Happiness'.[34] Additionally, the phenomenal world provides cues to virtuous and vicious behaviour through the sensations of pleasure and pain. Richard Haven contends that the 'world reveals the qualities of God, according to Hartley, and through the operation of psychological laws' enables humans to know him.[35] Hartley envisions the world as a system of

[30] Getty L. Lustila, 'John Gay and the Birth of Utilitarianism', *Utilitas*, 30.1 (2018), pp. 86–106 (n. 18), doi:10.1017/S0953820817000115.

[31] John Gay, *Preliminary Dissertation Concerning the Fundamental Principle of Virtue or Morality* (W. Thurlbourn, 1731), p. xix.

[32] Joseph Priestley, *Institutes of Natural and Revealed Religion*, 2nd edn, 2 vols (Pearson and Rollason for J. Johnson, 1782), I, p. 19, Eighteenth Century Collections Online <https://link.gale.com/apps/doc/CW0122282972/ECCO?sid=bookmark-ECCO&xid=8d97510e&pg=1> [accessed 23 May 2023].

[33] Gay, *Dissertation*, p. xii.

[34] Hartley, *Observations*, II, p. 22.

[35] Richard Haven, 'Coleridge, Hartley, and the Mystics', *Journal of the History of Ideas*, 20.4 (1959), pp. 477–494 (p. 483).

interdependent parts that function together to promote reciprocity between God and humankind via the mechanism of human cognition.

Hartley's commitment to relationality is evident in nearly every aspect of this theory, from the mutual relations that he envisions between God and humankind, to his understanding of physics, to his theory of cognition. Hartley, along with many others, thought that God created the world because he is 'a Being of great Benevolence', but he differs in that he sees the creation as mutually beneficial to both created and creator.[36] King asserts that 'the Communication of his Goodness' is one of 'the Ends for which the World is fram'd', but his exercise of benevolence is for humankind alone because God is perfect and 'can neither be profited nor incommoded by his Works'.[37] In contrast, Hartley believes that the fall affected God as well as humankind. He thinks that God suffers as he 'look[s] down upon Mankind passing through a Mixture of Pleasure and Pains' instead of enjoying the unmixed pleasures that he intended for them.[38] Furthermore, Hartley contends that the purpose of creation is mutual contemplation between God and humans, which cannot happen as it should in the postlapsarian world. Humans' worship is partial at best because they no longer have access to knowledge of his unadulterated being, but after the second coming 'the greater and more pure will his sympathetic Pleasure be, which arises from the Contemplation of Man' because it will once again be perfectly mutual.[39] The relationship Hartley envisions between humans and God is not unidirectional but characterized by mutual adoration. This relationship is fundamental for Hartley, and it directs his understanding of the phenomenal world.

Although Hartley does not address physics explicitly, it is clear that he thinks the universe is made up of Newtonian forces that structure it as a set of relations. Citing Newton as an inspiration for *Observations*, he uses the concept of aether discussed in *Mathematical Principles*. A dualist who sees matter as passive, Newton wonders how a material, albeit subtle, force like gravity functions despite its lack of self-initiating power. To solve this problem, he proposes an obscure substance that he calls aether to mediate material and immaterial substances. According to Smith, this 'subtle, elastic medium, pervades the whole of space' for Newton 'and was responsible for the transfer of forces from one locality to another'.[40] In the third book of *Opticks*, Newton poses questions and drops hints about how vibrations and aether

[36] Hartley, *Observations*, II, p. 28.

[37] William King, *An Essay on the Origin of Evil*, trans. by Edmund Law (W. Thurlbourn, 1731), pp. 53, 52.

[38] Hartley, *Observations*, II, p. 28.

[39] Hartley, *Observations*, II, p. 28.

[40] C.U.M. Smith, 'David Hartley's Newtonian Neuropsychology', *History of Behavioral Sciences*, 23.2 (1987), pp. 123–136 (p. 125).

might function together in cognition. He does not fully develop these ideas, but Hartley does, asserting that the 'Principles of Attraction and Repulsion … and of Vibrations, are dependent upon, and involved within each other'.[41] He sees forces such as gravity not only as an interplay of matter and motion that structures the phenomenal world but as entities that are so intertwined that it is impossible to make sense of them independently from each other. He states that 'Each Part, Faculty, Principle, *&c* when considered and pursued sufficiently seem to extend into the Boundaries of the others, and, as it were, to inclose and comprehend them all'.[42] These relationships among fundamental physical forces are not only 'agreeable to the Tenor of Nature' but also drive Hartley's theory of cognition, which relies on reciprocity among all parts of the system.[43]

King and Gay also influenced Hartley's thinking about relationships, especially as it pertains to the development of moral sense. King argues against the Manichean doctrine championed by Pierre Bayle, who could not reconcile evil with the existence of a purely benevolent God. According to this doctrine, the two principles of good and evil structure the universe, a contention that thinkers such as King and Gottfried Wilhelm Leibniz thought heretical because it implies the existence of more than one absolute force that rules all.[44] King argues that evil, whether natural evils such as hunger or illness or moral evils that result from freewill, inheres in matter. He claims that even though '*Good* and *Evil* are Opposites, [they] arise from the relations which things have to each other', and neither obtains on its own because they can only be judged in relationship to the other.[45] Matter entails evil, according to King, because it is divisible. Although imperfect, division necessitates relationships among the parts of the world, which stands in contrast to God who is indivisible and therefore epitomizes good. In King's view, the world cannot be free of evil, as it is a condition of creation. Gay, on the other hand, thinks that virtue depends on difference and divisibility. He asserts that 'Virtue, among all, or most, does imply some relation to *others*: where *Self* only is concern'd, a Man may be *prudent* but not virtuous'.[46] From Gay's perspective, virtuous action occurs only when it is directed towards and benefits a person other than oneself. This implies the necessity of differentiation because without different

[41] Hartley, *Observations*, I, p. 28.

[42] Hartley, *Observations*, I, p. 29.

[43] Hartley, *Observations*, I, p. 29.

[44] King influenced Leibniz's response to Bayle in *Théodicée* (1710). Where both men agreed that 'this world is the best of all possible worlds' that God could have created, they disagreed about freewill, with King contending that 'evil is consistent with the character of God'. R.R. Hartford, 'William King, the Man and His Work', *Theology*, 48.297 (1945), pp. 55–60 (p. 59).

[45] King, *Essay*, p. 73.

[46] Gay, *Dissertation*, p. xvii.

people, virtue would be impossible. Thus, the commonality across Hartley's intellectual influences is the emphasis on relationships, a concept that grounds his own system.

Hartley's form, content, and rhetoric provide evidence of his philosophical commitment to relationality. In addition to the concepts of aether and vibrations, Hartley borrowed Newton's method of theoretical reasoning. However, where Newton relied on experimentation to validate his hypotheses to the extent possible, Hartley combines analogy, mathematical demonstration, and observation from his medical practice, a method that relies on the 'interaction and complementary action of the proposition[s]'.[47] Hartley employs a reciprocal style of argumentation in which he uses established postulates as proofs of the ones that follow. He works in a 'circular fashion within the material rather than ... work[ing] through it from beginning to end, from premise to conclusion'.[48] The organization of his argument models both the relationality that structures cognition and the recursion that characterizes moral development in his account of them. Rhetorically, Hartley performs mutual interdependence, implying that, like Newton's forces, the physical, physiological, and theological aspects of his theory are so interconnected that they cannot be teased apart. He explains that he must choose an arbitrary starting point to begin his treatise because 'These Subjects are so much involved in each other that it is difficult, or even impossible, to begin any-where upon clear Ground'.[49] According to Charles I. Armstrong, in trying to reconcile 'the question of the whole' with the parts that comprise it, organicists such as Immanuel Kant and Friedrich Schelling struggled to resolve the competing claims of interdependence and hierarchy.[50] However, to prioritize one over the other, especially in a process-based model of organicism such as Hartley's, threatens a conceptual rupture. Hartley's admission that he cannot find 'clear Ground' on which to begin suggests that the very notion of a starting point undermines the unity of his system. Thus, his rhetorical style performs the conceptual work that he theorizes in *Observations*.

[47] Barbara Bowen Oberg, 'The Progress Toward the Perfection of Man: David Hartley and the Association of Ideas' (unpublished doctoral dissertation, University of California, 1973), p. 38.

[48] Oberg, 'The Progress Toward the Perfection of Man', p. 40; see Oberg, pp. 31–42 for a detailed discussion of Hartley's method, including examples of his 'mutually dependent' propositions (p. 39).

[49] Hartley, *Observations*, I, p. iv.

[50] Armstrong, *Romantic Organicism*, pp. 13–14.

Cognitive Organicism

The mutual dependence manifest in Hartley's method and rhetoric structures his theory of cognition, which advances a model of organic holism that attempts to account for association scientifically and physiologically and to explain how it facilitates the connection between God and his creation. Describing his project in the first volume of *Observations* as an investigation into 'Sensations, Ideas, Motions, and their mutual Influences and Relations', Hartley indicates through the phrase 'mutual Influences and Relations' that he sees cognition as a reciprocal process.[51] Because of his belief that God created the world so that he and his creatures could commune, Hartley needed to elucidate how God makes himself accessible to humankind. He postulates that God designed nature and the body to work together to mediate his radical alterity by making his infinitude discernible by finite beings.[52] Hartley states that the 'Course of the World, and the Frame of our Natures, are so ordered, and so adapted to each other, as to enforce Benevolence upon us'.[53] The phrase 'enforce ... upon us' suggests that embodied interaction with the external world inevitably leads to knowledge of God. The reciprocity between humankind and nature ensures communion between human beings and God. Hence, the relationship between humans and nature manifests organicism because together they create something greater than themselves—knowledge of God.

Since, for Hartley, original sin created an epistemological rather than ontological break, the 'problem was not a metaphysical one, but only a problem of knowing'.[54] Thus, cognition plays a crucial role in Hartley's organicism under both pre- and postlapsarian conditions because it facilitates knowledge of God. However, because this knowledge was distorted by the fall, human beings became subject to association, the process by which repeated experience forms mental and emotional connections between concepts, events, and objects, and which directs all human behaviour. Association is crucial in the fallen world because it determines whether a person will move closer to or further away from God. Although Locke developed Aristotle's theory of association in *Human Understanding*, Hartley adapts it by combining it with Gay's ideas about virtue and vice. As the 'first modern philosopher to maintain that happiness-promotion is the aim of moral action' (and therefore 'the first utilitarian'), Gay was also the first to apply Locke's ideas to a moral rather than

[51] Hartley, *Observations*, I, p. 33.

[52] See Emmanuel Levinas, *Of God Who Comes to Mind*, trans. by Bettina Bergo (Stanford University Press, 1998), in which Levinas uses the phrase 'radical alterity' to explore the origins of humankind's concept of God.

[53] Hartley, *Observations*, II, p. 16.

[54] Oberg, 'The Progress Toward the Perfection of Man', p. 122.

an epistemological context.[55] Gay argues that 'Ideas themselves about which Morality is chiefly conversant ... are all *mixed Modes*, or compound Ideas arbitrarily put together, having at first no Archetype or Original existing'.[56] Compound ideas refer to abstractions, which, according to Locke and Gay, have no correspondence to either Platonic ideals or physical objects in the way that simple ideas such as rocks or trees do. Rather, simple ideas form abstract ideas when they combine together as a person moves through and experiences the world. When applied to association, these ideas can become connected arbitrarily, for better or for worse. However, 'Gay does not say anything concrete about promoting happiness' but 'instead provides a structure through which to approach these questions', which Hartley uses to develop his account of how experience fosters personal dispositions towards virtue or vice by considering association in both the moral and epistemological contexts.[57] Association begins with cognition, which, according to Hartley, influences how clearly we apprehend God's truth and how well we can commune with him in the fallen world.

The cognitive processes that Hartley describes in the first volume of *Observations* exemplify the principles of organicism. Dependent on the interactive relationships among multiple interdependent parts, cognition is the mechanism by which his organicism works. To fill in the details missing from Gay's work, Hartley goes directly to Locke, drawing on tenets from his sensationalist psychology which he supplements with natural religion to posit cognition as a series of processes that are contingent upon structural compatibility between nature and the human body. Although Locke claims that no one is born with innate knowledge and that all ideas originate in sensation, he says nothing about the mechanisms that make this possible. The relationship that he hypothesizes between sensation and ideation strongly implies an interdependent relationship between body and world, yet Locke actively eschews pursuing this line of enquiry. On the very first page of *Human Understanding*, he distances himself from the problem of Cartesian interaction, declaring that 'I shall not at present meddle with the physical consideration of the mind, or trouble myself to examine wherein its essence consists, or by what motions of our spirits or alterations of our bodies, we come to have any sensation by our organs, or any ideas in our understandings'.[58] Hartley combines Locke's ideas with the natural theological tenet that God created the world to promote human happiness. While thinkers such as Priestley saw a fit between nature and

[55] Lustila, 'Gay and Utilitarianism', p. 87.

[56] Gay, *Dissertation*, p. xii.

[57] Lustila, 'Gay and Utilitarianism', p. 94.

[58] Locke, *Human Understanding*, p. 1.

humankind that demonstrates this belief, Hartley traces it out in his account of how they function together to facilitate knowledge of God.

Tackling the hard problems that these early thinkers shied away from, Hartley situates the mind as the intersection of the body and its surround. He posits that the body and the external world actively cooperate in the process of gathering and transferring sense data. At its most basic, Hartley's theory posits that vibrations transmit these data from the world to the body, which processes them into embodied knowledge. Although Hartley's 'model is structurally simple, it is dynamically complex', positing cognition not as a static process wherein passive percipients take in and reconstruct the world on the canvas of the mind, but rather as ceaselessly active and interactive.[59] Scholars have mistakenly interpreted his process of data transfer from the external world as unidirectional, with objects of sense impressing or inscribing information on the body-mind, but, in fact, both the body and the world are actively involved. The body initiates the process by stimulating objects to produce the vibrations that carry sensory information, which '*are excited, propagated, and kept up, partly by the Aether ... and partly by ... the active Powers of the medullary Substance of the Brain, Spinal Marrow, and Nerves*'.[60] Here, Hartley clearly states that the nervous system has 'active Powers', which bears out in his explanation of how the body stimulates and processes vibrations.

Hartley envisions the information transfer as a feedback loop of constant communication between the embodied mind and the external world. Vibrations do not simply radiate off objects of sense, waiting to impress themselves on passers-by. Rather, they arise from the joint action that occurs between the objects, nerves, and the aether (the medium by which vibrations travel). In Hartley's example, the hand's remarkable motility allows people to 'rub the Ends of the Fingers against the tangible object' so that the 'Friction may, by exciting Vibrations, and a consequent Contraction in the certain muscular Fibrils belonging to the *Papillae*, distend and erect these, and thereby increase their Sensibility'.[61] In this twofold process, the physiological structure of the hand enables it to manipulate the object to increase its vibrational intensity so that it produces more waves to generate more data, making it easier to perceive.[62] At the same time, the friction increases the fingers' sensitivity so that they are better able to receive vibrations. In Hartley's cognitive system, bodily and environmental structures interact dynamically to provide

[59] Allen, *David Hartley*, p. 401.

[60] Hartley, *Observations*, I, p. 13.

[61] Hartley, *Observations*, I, p. 116.

[62] Smith notes that Hartley did not think of vibrations as atoms bumping into each other but as 'unit waves: Fourier elements, not billiard balls!' Smith, 'Newtonian Neuropsychology', p. 130.

the embodied mind with the data necessary to begin understanding nature's true purpose.

Hartley's account of colour perception provides another example of the active relationship between the world and the body. Rather than passively receiving vibrations, when light waves contact the eye the organs of sense begin processing and organizing sense data into knowledge before transmitting them through the nervous system, and, in the process, changing them. Hartley asserts that each colour in the spectrum has its own frequency that affects the retina differently, and, conversely, the retina processes each colour in a different way. For example, he explains that 'Red is less refracted by the Coats and Humours of the Eye than Violet'.[63] This means that when light waves of 430 trillion hertz (red) move through the clear surface of the eye ('Coats') and intraocular fluid ('Humours'), it produces a different effect than those of 750 trillion hertz (violet), which causes them to appear different from each other.[64] However, the eye also alters the frequencies once they enter the retina. Red, says Hartley, 'will not have its Intervals so much diminished in Proportion; whence they may be to those of Violet … at their arrival at the Retina'.[65] Thus, as the frequencies of light waves interact with the structure of the eye, they produce visual distinctions between the colours that contributes to a percipient's colour perception. Nor is the eye unique in its ability to modify sensory stimuli; the 'physical vibrations of the air are [also] refined and transformed in their progress through the inner and outer ear'.[66] In this way, the sense organs adapt the information they gather from the external world into a form that the body can use.

This compatibility between the world and the body further demonstrates the reciprocity between parts that characterizes Hartley's organicism. According to Hartley, specific vibrations are especially adapted to individual sense organs and to specific regions of the brain, depending on the stimulus modality. Knowledge of the whole depends on the specialization of parts, that is, both the organs of perception and those found in nature. The structure of body parts such as the ear equips them to process particular vibrations fitted to their functions. The outer ear and the ear canal, for example, are 'cartilaginous, and seem by this means to be peculiarly fitted for receiving and retaining the Vibrations of the Air, and for communicating them to the *Membrana Tympani*', the eardrum.[67] The shape of the ear as well as its constitutive tissue makes it aptly suited to processing sound vibrations. Similarly, the eyes are structured to

[63] Hartley, *Observations*, I, p. 192.

[64] William Harris and Craig Freudenrich, 'How Light Works', *HowStuffWorks*, 1 July 1970 <https://science.howstuffworks.com/light.htm> [accessed 31 May 2019].

[65] Hartley, *Observations*, I, p. 193.

[66] Allen, *David Hartley*, p. 112.

[67] Hartley, *Observations*, I, p. 223.

direct stimulation.[79] As the sense data linger in the nervous system, the body processes them into miniature vibrations, that is, vibratiuncles, that permanently alter the tenor of the nervous system by repeatedly displacing the nerves from their resting state and changing the way the nerves vibrate.[80] Vibratiuncles 'underlie memories and ideas and motor patterns', accruing in associative 'trains' that settle into 'habitual patterns of thinking, feeling, perceiving, and acting, ... such as confidence, avarice, laziness, worry', and so forth.[81] Thus, Hartley develops Gay's treatise on virtue and vice, using association to account for why some people develop the former while others develop the latter.

Building on the connection between virtue/vice and sensation/emotion, Hartley develops the implications of Gay's account, which assigns moral value to sensations while also making embodied experience the starting point of morality. Gay claims '(following Hobbes and Locke) that we use "good" to refer to objects that produce pleasure, and "bad" to objects that produce pain' and that 'there is a correspondent passion directed at the object: love in the case of goodness; hatred in the case of badness'.[82] Hartley also equates pleasure and pain with the two basic emotions love and hate, from which all other emotions stem. He does not, however, consider how these passions are directed outwards towards the objects that inspired them. Rather, Hartley turns inwards towards character formation to explain how moral sense, or conscience, develops. He claims that 'all the mere Sensations were, in their original State, either Pleasures or Pains' or some combination of the two.[83] Seeing pleasure and pain as the polar ends of a continuum, Hartley calls the range between the two the 'Solution of Continuity', which helps him explain why and how some people become acclimated to vice, even though it was generally believed to result in painful outcomes in the long run.[84] This paradigm not only puts pleasure

[79] Smith, 'Newtonian Neuropsychology', p. 127.

[80] Smith, 'Newtonian Neuropsychology', p. 127. For a detailed account of this process, see Smith, pp. 127–129. The term vibratiuncles is a solecism of the Latin *vibraituncula*, or miniature. Allen, *David Hartley*, p. 117. Allen notes that 'if pushed into a position often enough, an elastic body will eventually learn to occupy that position', which is hypothetically what happens in Hartley's theory. Allen, *David Hartley*, p. 120.

[81] Oldfield and Oldfield, 'Hartley's *Observations*', p. 372; Varela, Thompson, and Rosch, *The Embodied Mind*, p. 67.

[82] Lustila, 'Gay and Utilitarianism', p. 98.

[83] Hartley, *Observations*, I, pp. 137, 41.

[84] Hartley, *Observations*, I, p. 36. Seeing pleasure and pain as 'the same species' of phenomenon that vary only by 'a mere Difference in Degree', Hartley adopts the term the solution of continuity from medicine to explain how 'an agreeable Warmth may be made to pass into a troublesome, or burning Heat, by Increase, or Continuance' or how something painful becomes tolerable, even pleasurable, after repeated exposure. Hartley, *Observations*, I, p. 35. Ironically, in contemporary medicine a solution of continuity referred to a rupture in or separation of connected physical structures, as in a fractured limb.

and pain on a relative scale, but it also assumes that every physical sensation, even a neutral one, falls somewhere along the continuum and, therefore, has an emotional valence, even when it is not consciously registered.[85]

Hartley's theory posits virtue and vice not just as intellectual orientations but also as states of being inflected by emotion. While Hutcheson also saw moral sense in terms of sentiment, which scholars interpret as feeling, in opposition to rationality, in Hartley's theory emotions are embodied states that develop over time. In his model of organicism, the structures of thought and feeling that function interdependently to produce knowledge manifest tangibly in the body as affective states that underwrite all human attitudes and behaviour. Hartley's theory challenges the Cartesian *cogito*, not only by tying cognition and morality to embodiment and embeddedness but also by troubling the dichotomy between ratiocination and the emotions, which generally have been viewed as the domains of two discrete faculties. Although Thomas Dixon, a historian of emotions, warns that in the eighteenth century the 'reason-passion dichotomy was not so stark' as most scholars of the period have assumed, Hartley's account of neural transduction, the process by which sensations become ideas, is striking because it unites two aspects of the human psyche into one system: thought and feeling.[86] Dixon draws on definitions of terms such as passions and affections found in Samuel Johnson's 1755 dictionary, which reveal much overlap. However, in contradistinction to Dixon's claim, the entries for 'thought' and 'feeling' have binary definitions for the terms, with the former referring to an 'operation of the mind; the act of thinking' and the latter referring to the 'sense of touch', 'Sensibility', and 'Perception'.[87] For Hartley, feeling is neither separate from nor an epiphenomenon of cognition, but rather the consequence of an embodied and embedded mind that cannot transcend the phenomenal world. Joining thought and feeling in one process of association highlights the unity that Hartley saw among the parts of a system generally considered unconnected.

The term feeling unites the thinking mind and the sensitive body in Hartley's theory because he does not put them into a cause-and-effect relationship in which people have thoughts first and then have feelings about them, or vice versa. Instead, both originate in stimuli from the external world, which puts them on equal footing. He asserts that 'Sensation is the common Foundation

[85] Hartley acknowledges that 'there are many Ideas, i.e., internal Feelings, which have no Names'. Hartley, *Observations*, I, p. 211. I call these nameless 'internal feelings' neutral sensations.

[86] Thomas Dixon, *From Passions to Emotions: The Creation of a Secular Psychological Category* (Cambridge University Press, 2003), p. 17.

[87] Samuel Johnson, *A Dictionary of the English Language* (J. and P. Knapton; T. and T. Longman; C. Hitch and L. Hawes; A. Millar; and R. and J. Dodsley, 1755), n.p., Internet Archive <http://archive.org/details/johnsons_dictionary_1755> [accessed 30 May 2024].

of all these' mental phenomena.[88] Juxtaposing thought against sensation rather than feeling, Hartley uses the term consistently throughout *Observations* to refer to ideation, in accordance with Johnson's definition. However, thought serves as an umbrella term for '*Intellectual Ideas*' and 'Ideas of Feeling'.[89] Indeed, Hartley's decision to pair both 'intellectual' and 'feeling' with 'ideas' indicates that he sees both as a form of ideation. Additionally, he employs the phrase 'Sensation, Thought, and Motion' as a synecdochic shorthand to capture each stage in the cognitive cycle, which further demonstrates that thought incorporates the intellectual *and* the emotional aspects of cognition.[90] Interestingly, while both are products of sensibility, which Johnson defines as 'quickness of sensation', neither thought nor feeling ends with perception.[91]

For Hartley, feeling functions as a versatile concept theoretically and rhetorically. In the first volume of *Observations*, Hartley uses feeling in two senses, first to refer to a particular stimulus modality and second to refer to the formation of ideas and concepts. The phrase 'Sense of Feeling' in Hartley's theory denotes the sense of touch, which leads to the sensations that attend it and connects to the second usage.[92] Once sensations enter the nervous system through the organs of sense, they become 'internal Feelings of the Mind which', according to Hartley, 'arise from the Impressions made by external Objects upon the Several Parts of our Bodies'.[93] These 'feelings of the mind' do not come just from touch but include sensations generated by all other sensory modalities as well. They also provide the raw material for all ideas. Just as aether bridges the inner world of human thought and the outer world of objects, sensation acts as a lynchpin between external stimuli and higher-level cognitive processing that enables the embodied mind to internalize external objects into thoughts and feelings. In keeping with Locke's distinction between simple and compound ideas, Hartley separates 'internal Feelings of the Mind' into two corresponding classes.[94] '*Ideas of Sensation*', like Locke's simple ideas, have direct external referents, such as cold or blue; '*Intellectual Ideas*' that reference concepts, such as justice or benevolence, parallel compound ideas.[95] The latter consists of clusters of the former, which means that all intellectual ideas contain some element of feeling. For example, Hartley claims that certain words, such as

[88] Hartley, *Observations*, I, p. 368.

[89] Hartley, *Observations*, I, pp. ii, 137.

[90] Hartley, *Observations*, I, p. 32.

[91] Johnson, *Dictionary*, n.p.

[92] Hartley, *Observations*, I, p. 117.

[93] Hartley, *Observations*, I, p. ii.

[94] While Hartley accepts these Lockean distinctions, he rejects Locke's theory of reflection. He states 'that all the most complex Ideas arise from Sensation; and that Reflection is not a distinct *Source*, as Mr. *Locke* makes it'. Hartley, *Observations*, I, p. 360.

[95] Hartley, *Observations*, I, p. ii.

'Algebraic Quantities', 'scientific Terms of Art, and the most abstract general Terms' can be defined only with reference to other words, making them highly abstract, yet he ventures that 'mental Emotions are apt to attend some of these'.[96] Conversely, he claims that 'our Passions or Affections can be no more than Aggregates of simple Ideas united by Association'.[97] In the latter case, he considers feeling states that generally relate to the appetites as intellectual ideas; in the former, decidedly abstract concepts, typically viewed as the province of reason, have some emotion inherent in them because all ideas originate in sensation.

The role Hartley assigns to 'feeling' in even abstract cognitive processes is critical because it reveals the central role emotion plays in cognition and association, from which moral sense emerges. Hartley's theory of moral development illustrates the way his organicism functions not just as a structuring concept but also as a series of embodied experiences that make up what Hartley calls 'personal Identity', by which he means the self.[98] Often Hartley associates these feelings with spending time in nature or with art, feeding the imagination. In this way, he connects morality with aesthetics, and having set up cognition as the mechanism of moral growth he moves on to a detailed explanation of how the self develops by passing through the 'intellectual Affections'.[99] Here, moral sense, according to Hartley, develops from embodied interactions that create affective states of being that, under the right circumstances, lead individuals closer to God. In the second volume of *Observations*, the various senses of feeling coalesce into something like Eugene T. Gendlin's notion of 'felt sense', which he describes as deeply embodied sensations that occur prior to linguistic articulation, or 'how we feel *our bodies from the inside*'.[100] Felt sense approximates Hartley's ideas about how human beings come to know God and the truth of his benevolent omnipotence. Where traditional models of aesthetic organicism tend to address how artworks fit together or how they express a reality that transcends the phenomenal world, Hartley's organicism sees aesthetics as crucial to developing virtue and reaching the highest stages of moral growth.[101]

[96] Hartley, *Observations*, I, p. 278.

[97] Hartley, *Observations*, I, p. 368.

[98] Hartley, *Observations*, I, p. 399. Allen explains that, for Hartley, the 'self' is not a 'substance or a thing' but an active, ever-changing 'complex of complexes, a latticed molecular compound of associations, a hypercomplex idea'. Allen, *David Hartley*, p. 333.

[99] Hartley, *Observations*, I, p. 368.

[100] Eugene T. Gendlin, 'Three Assertions About the Body', *Folio*, 12.1 (1993), pp. 21–33 (p. 22).

[101] See Devin Zane Shaw, 'The "Keystone" of the System: Schelling's Philosophy of Art', in *The Palgrave Handbook of German Idealism*, ed. by Matthew C. Altman, Palgrave Handbooks (Palgrave Macmillan, 2014), pp. 518–538.

Aesthetics and Moral Sense

In the postlapsarian world, Hartley's account of moral development is expressive of his cognitive organicism not least because it is the means of becoming more aware of God's truth and growing closer to him. For Hartley, the stages of moral growth, or intellectual affections, close the epistemological circle that was broken in humankind's first disobedience. Here, the intellectual affections not only represent another subsystem structured by the principle of relationality but, like the interactions of the physiological body and the external world, the recursive relationship between the intellectual affections and sensation expresses Hartley's cognitive organicism. The affections consist of embodied affective states of being that result from cognition. Grounded in the imagination, the first stage of moral growth, they rely on embodied aesthetic experiences to develop. In contrast to Hutcheson, who saw moral sense as the capacity to favour long-term over short-term good, in *Observations* Hartley sees it as the culmination of a lifetime of experience: moving from imagination to ambition, self-interest, sympathy, and theopathy, and culminating in moral sense, the intellectual affections develop as people mature, though there is no guarantee that everyone will reach the highest stage as much depends on circumstance and effort. Furthermore, where thinkers such as Butler and, later, Adam Smith, believed that enlightened self-interest motivated moral behaviour, Hartley sees the final stage of moral development as a form of self-annihilation as people become subsumed by the love of God and other human beings, which is consistent with the ultimate aim of moral development—to restore, to the extent possible, individuals, and humankind more generally, to the knowledge of and communion with God that they enjoyed before the fall.

Although Hartley envisions the intellectual affections as progressive, developing from infancy and throughout life, he also sees their maturation process as recursive. Not only must each one develop sufficiently for the subsequent ones to evolve, but the first set of affections must mature enough for the second set to develop. Adapting the early Christian bipartite model advanced by Augustine and Thomas of Aquinas that associates the 'passions and appetites' with the 'lower animal soul' and 'the affections' with the 'higher rational soul', Hartley divides the six intellectual affections into lower and higher.[102] In contrast to Butler, who also distinguishes between two categories of moral motivation—the 'particular appetites, passions, and affections' and 'superior principles'—but fails to define them clearly, Hartley sets out a clearly delineated taxonomy.[103]

[102] Dixon, *From Passions to Emotions*, p. 21.

[103] David McNaughton, 'Butler on Benevolence', in *Joseph Butler's Moral and Religious Thought: Tercentenary Essays*, ed. by Christopher Cunliffe (Clarendon Press, 1992),

He categorizes imagination, ambition, and self-interest as lower affections, while sympathy, theopathy, and moral sense belong to the higher class. Each one develops from and is a compound of the others, so that once 'Sensation generate[s] Imagination then will Sensation and Imagination together generate Ambition; Sensation, Imagination, and Ambition, Self-interest' and so forth, 'Till at last, by the numerous reciprocal Influences of all these upon each other, the Passions arrive at that Degree of Complexness, which is observed in fact'.[104] Like the contemporary concept of recursion, under 'reciprocal Influences' each successive affection is determined by and emerges from the preceding ones in accordance with the 'Rule of Life', which Hartley, following Gay, defines as 'Compliance with the Will of God'.[105] After the lower affections have matured enough for the higher ones to develop, the latter recursively guide and encourage the continued progress of the lower three.

Hartley's theory of intellectual affections proposes a developmental model that originates in sensation and reaches its peak in as pure a love of God as possible. In the prelapsarian state, people did not need to pass through stages because interacting with nature automatically enacted embodied knowledge of God. However, in the postlapsarian world not only do human beings need to cultivate knowledge of him, but they also have to contend with the errors introduced by the fall. Consequently, in accordance with the doctrine of association, the intellectual affections develop unevenly, depending on circumstances and environment. While they typically 'succeed in order from Infancy to old Age', people are subject to all sorts of influences that can hinder progress or lead them astray, and it is possible to get stuck.[106] When people allow one affection to take precedence over the others, with the exception of moral sense, that affection dominates. For example, people who do not mature much beyond ambition or self-interest may be dominated by the desire for praise or the drive for wealth, and those stuck in theopathy may be prey to enthusiasm or superstition. In contrast to Butler, who sees 'the present World' as 'a State of Discipline', full of 'Snares and Temptations of Vice' that 'render being upon our Guard, Resolution, and the Denial of our Passions, necessary', Hartley does not think that the world is meant to ensnare people.[107] Rather, he optimistically posits that once science figures out how association works, people will be able to purge the errors of perception. He suggests that if humankind could master the mechanics of character formation 'by reversing the steps of ...

pp. 269–291 (p. 269). According to McNaughton, Butler's 'failure to distinguish' his categories 'carefully has confused the commentators and, perhaps, even Butler himself'. McNaughton, p. 270.

[104] Hartley, *Observations*, I, pp. 368–369.

[105] Hartley, *Observations*, II, p. 196.

[106] Hartley, *Observations*, II, p. 419.

[107] Butler, *Analogy*, p. 96.

Association', humankind could 'learn how to cherish and improve good ones [and] check and root out such as are mischievous and immoral'.[108] Hartley's goal is to repair, to the extent possible, the epistemological damage done by the fall. When properly directed, he asserts, association 'has a Tendency to reduce the State of those who have eaten of the Tree of Knowledge of Good and Evil, back again to a paradisiacal one'.[109] However, for Hartley, this process begins with the first intellectual affection, imagination, which is the source of knowledge.

Early scholars dismissed Hartley's account of the imagination because he 'assigned the esemplastic power so low a place in the hierarchy of pleasures'.[110] Yet, as the first intellectual affection, it provides the base that all others incorporate and build upon. It is crucial to the higher affections and, therefore, the highest stages of moral development. Here, Hartley's understanding of the imagination is similar to Samuel Taylor Coleridge's definition in the *Biographia Literaria* (1817). Coleridge holds 'it to be the living power and prime agent of all human perception'.[111] Like Coleridge's primary imagination, Hartley's first intellectual affection emerges from sensations that arise from embodied interaction with the world. It connects people to God's creation by giving them a sense of their surroundings. As one of the first three intellectual affections, imagination is initially oriented towards the self. Considered developmentally, infants and children must first get a sense of themselves before they can become aware of the social world and sympathize with others.

The transition to the first higher intellectual affection, sympathy, requires imagination. Along with ambition, which is the desire to perform well and be admired by others, imagination is part of gross self-interest. Hartley subdivides self-interest into three stages, which, in addition to gross, includes refined and rational self-interest. The first stage centres on 'the Pursuit of the Means for obtaining the Pleasures of Sensation, Imagination, and Ambition', while rational self-interest marks the turning point from the lower to the higher affections as people begin to realize that their long-term personal good is bound up with the social good.[112] Similar to Adam Smith's better-known but later account of sympathy in *Theory of Moral Sentiments* (1759), Hartley contends that the 'very appearance and Idea of any Kind of Misery which they [people] have experienced ... raise[s] up in their nervous Systems a State of Misery ...

108 Hartley, *Observations*, I, p. 81.

109 Hartley, *Observations*, I, p. 83.

110 Hoxie N. Fairchild, 'Hartley, Pistorius, and Coleridge', *PMLA*, 62.4 (1947), pp. 1010–1021 (p. 1021).

111 Samuel Taylor Coleridge, *Biographia Literaria, or, Biographical Sketches of My Literary Life and Opinions*, ed. by James Engell and W. Jackson Bate, Collected Works of Samuel Taylor Coleridge, 7, 2 vols (Princeton University Press, 1983), I, p. 304.

112 Hartley, *Observations*, II, p. 271.

on account of their Imaginations'.[113] Although sympathy is stronger when one has experienced a similar experience, it nonetheless relies on the ability to imagine oneself in another person's situation. Thus, Hartley's definition of sympathy reveals that the imagination is an embodied faculty that enables people to feel what others feel.

Hartley also considers the imagination the faculty of aesthetic appreciation, which facilitates embodied affective responses to art and nature. Although 'the Pleasures of the Imagination were not intended for our primary Pursuit', they are necessary to cultivate theopathy, which Hartley explicitly connects with the imagination, stating that the 'Pleasures of Imagination point to Devotion in a particular manner'.[114] His discussion of the pleasures and pains of imagination, again, resemble later discussions of aesthetics that are more familiar to Romantic scholars. Indeed, Hartley describes the reaction to a dangerously steep 'Precipice' or a looming 'Mountain of Snow' in terms similar to Edmund Burke's explanation of the sublime in *Philosophical Enquiry into the Origin of Our Ideas of the Sublime and Beautiful* (1757, 1759). Hartley states that 'Fear and Horror magnify and enliven all other Ideas and by degrees pass into Pleasures, by suggesting the Security from Pain'.[115] On the other hand, he associates beauty with the 'pleasant Tastes, and Smells, and the fine Colours of Fruits and Flowers, the Melody of Birds', which are pleasurable in their own right.[116] The principal theopathic pain, fear of God, closely resembles the experience of the sublime, and when it is 'tempered with Love', the chief pleasure of theopathy, it 'become[s] Awe, Veneration, and Reverence'.[117] Even sexual attraction, which Hartley, like Burke, attributes to the influence of beauty, serves as 'an intermediate Step between the gross sensual [affections], and those of pure Esteem and Benevolence', which 'moderate, spiritualize, and improve the first, and ... are ultimately converted into' higher ones.[118] As a subset of cognitive structures particularly attuned to art and nature, the imagination can be understood as a Romantic structure of engagement that has access to God's truth embodied in the phenomenal world; it aids in 'the Contemplation of God, and his Attributes, and of our Relation to Him'.[119] It must, however, be guided by theopathy in order to move beyond the appreciation of art and nature for their own sakes and to culminate in moral sense.

Positing an especial affinity between the imagination and moral sense, therefore, Hartley links aesthetics and religious feeling to the highest stage of

[113] Hartley, *Observations*, I, pp. 474–475.
[114] Hartley, *Observations*, II, pp. 244, 246.
[115] Hartley, *Observations*, I, p. 419.
[116] Hartley, *Observations*, I, p. 419.
[117] Hartley, *Observations*, I, p. 491.
[118] Hartley, *Observations*, I, p. 435.
[119] Hartley, *Observations*, I, p. 486.

moral growth. The language of moral sense, he explains, 'is borrowed, in great measure, from the Pleasures and Pains of Imagination' because they are 'applicable to those of the Moral Sense with a peculiar Fitness and Significancy'.[120] The aesthetic pleasures of the imagination are crucial because they 'improve and exalt the Mind' and encourage feelings such as 'Love, Awe, and other exalted Emotions of Mind'.[121] Robert Brainard Pearsall explains that when 'original sensation combine[s] into imagination' it generates 'a sort of reflexive recreation of the original' that develops 'into a set of gradually rarefied inner experiences' that Hartley associates with moral sense.[122] Hartley defines moral sense as the 'disinterested Love and Esteem for ... others' that guides human attitudes and actions, and moves people beyond the enlightened self-interest that moral philosophers such as Smith thought necessary to motivate virtue.[123]

In contrast to Butler, for example, who claims that '*Benevolence and the Pursuit of publick Good hath just the same Respect to Self-love and the Pursuit of private Good, with all other particular Passions*', Hartley sees selflessness as the objective of moral sense.[124] Because the intellectual affections are recursive, 'though it be impossible to ... proceed without the other intermediate Principles, and particularly that of rational Self-interest', says Hartley, 'we ought never to be satisfied with ourselves, till we arrive at perfect Self-annihilation, and the pure Love of God'.[125] A state of love for God, his creation, and other people, freed from self-interest, self-annihilation 'is not a transcendent state of amnesia—an experience whereby one's sense of being a person is annulled'.[126] Neither does it involve rejecting the phenomenal world. Rather, to achieve self-annihilation, people must continually replace selfish motives with those directed towards benevolent actions that promote other people's happiness. Hartley's theory of moral sense thus depends on his theory of cognition, whereby an embodied mind embedded in the phenomenal world need not transcend it to achieve union with God. Motivated by social interest and virtue rather than hope for a reward or fear of punishment, people who have sufficiently developed their moral sense appreciate the pleasures of the imagination because they reveal God's truth in nature.

Without the imagination, it would be impossible for people to form an

[120] Hartley, *Observations*, I, p. 496; Hartley, *Observations*, II, p. 245.

[121] Hartley, *Observations*, II, pp. 254, 292.

[122] Robert Brainard Pearsall, 'Wordsworth Reworks His Hartley', *The Bulletin of the Rocky Mountain Modern Language Association*, 24.2 (1970), pp. 75–83 (p. 80).

[123] Hartley, *Observations*, I, p. 493.

[124] Joseph Butler, *Fifteen Sermons Preached at the Rolls Chapel* (Botham, 1726), p. 219.

[125] Hartley, *Observations*, II, p. 282.

[126] Allen, *David Hartley*, p. 332. See also Allen, pp. 332–334; Oberg, 'The Progress Toward the Perfection of Man', pp. 127–130; Fairchild, 'Hartley, Pistorius, and Coleridge', pp. 1019–1020.

idea of God, who by Hartley's account, is not directly perceptible in the postlapsarian world. However, he can be known, albeit imperfectly, through his creation. Thus, Hartley asserts that humans 'judge of the Divine Power and Knowledge by their Effects in the Constitution of the visible World'.[127] Children base their knowledge of God on the things they perceive with their senses, most particularly their caregivers. Furthermore, 'since many Actions and Attributes belonging to Men' are 'applied to God' when speaking about him in everyday language, 'Children in their first Attempts to decypher the Word God, will suppose it to stand for a Man' and transfer the embodied emotions associated with people onto him.[128] Viewing 'emotion [as] a fluid like electricity or water', Hartley theorizes that 'it jumps or flows from one experience, memory, word, or cluster of words to another', a process that transfers children's feelings to the abstract concept of God.[129] Here, Hartley, who got the idea of transference from Gay, flips it from negative to positive. Where Gay asserts that the 'Power of *Association* is so great not only to transport our Passions and Affections beyond their proper bounds, ... but also it is able to transfer them to improper Objects', Hartley applies this idea to knowledge of God.[130]

As children get older and increase their stock of ideas through association, they learn that God is omnipotent, omniscient, and omnipresent. These intangible abstractions contradict what they already know about the phenomenal world, which confuses them. When confronted with 'Inconsistencies and Impossibilities in their Ideas, they must feel great Perplexity of Imagination, when they endeavour to conceive and form definite Ideas' of God, and 'this Perplexity will add to the Vividness of the Ideas' surrounding him.[131] Rather than hindering knowledge of God, the confused imagination bridges the gap between what children already know and the abstract notion of him. It also adds a sense of wonder, mystery, and awe that leads to theopathy. Eventually and 'by degrees', the abstraction that characterizes God's nature 'destroys the Association of the fictitious visible Idea' of him.[132] Thus, human knowledge of God, in Hartley's account, consists of complex embodied feelings and ideas that people receive via the imagination, which serves an ethical function that is 'consistent with our *summum Bonum*' because its pleasures 'yield ... to more exalted and pure pleasures' associated with God.[133] Hartley's cognitive organicism therefore operates both as a model of the phenomenal world in

[127] Hartley, *Observations*, II, p. 13.
[128] Hartley, *Observations*, I, p. 486.
[129] Allen, *David Hartley*, p. 267.
[130] Gay, *Dissertation*, p. xxxi.
[131] Hartley, *Observations*, I, p. 487.
[132] Hartley, *Observations*, I, p. 487.
[133] Hartley, *Observations*, I, p. 422.

which all parts are connected and as an expression of the embodied, aesthetic experiences that enable knowledge and awareness of God as people move through the intellectual affections and stages of moral growth.

Universal Salvation and Freewill

Hartley formulated his theory of mind and moral development because he could not reconcile himself to the Cartesian-influenced notion that not only did God separate himself from his creation after the fall but he placed human beings in an inherently sinful environment that would tempt them into vice. Nor could he believe that a loving God could damn his creatures to eternal punishment. According to Oldfield and Oldfield, as a child Hartley had a superstitious, 'puritan fear of hell-fire', which made the doctrine of universal salvation particularly appealing.[134] For Hartley, 'hell does not exist somewhere "out there", [as] a supernatural realm analogous to a physical one', but rather he envisions it as a state of alienation from God.[135] Thus, he could not accept the implications of Cartesianism, which makes the entire universe, except the human mind, irrelevant or inimical to salvation by splitting the moral and physical aspects of human beings in two and pitting them against each other. Hartley's organicism authorizes universal salvation because he could not believe 'that the Fall had produced a permanent fragmentation of the universe into a realm of bliss and a realm of endless pain'.[136] Instead, the promise that original unity would someday be restored drives his theory of organicism.

Under organicism, the integrity of the whole depends on each constituent part, and if one part is removed the system breaks down. Because Hartley sees the fall as epistemological rather than ontological, the doctrine of eternal damnation becomes a logical fallacy. It reinforces the idea that God's world has been permanently riven by the fall because it implies that some portion of humankind will remain permanently separated from him. By the same logic, 'God would be incomplete to the extent that the whole is not recovered' because the 'completeness of the divine nature requires the completeness of human nature'.[137] The concept of divine retribution also dictates that the creation will remain in an imperfect state, lacking original wholeness, which further implies that God will also lack wholeness, an impossibility according to eighteenth-century theologians. Since Hartley saw the purpose of human life as continued movement towards pure knowledge of God, damned souls could not fit into this as they would be forever barred from achieving humankind's

[134] Oldfield and Oldfield, 'Hartley's *Observations*', p. 373.
[135] Allen, *David Hartley*, p. 326.
[136] Allen, *David Hartley*, p. 39.
[137] Allen, *David Hartley*, pp. 352–353.

divinely ordained teleology. Under the doctrine of universal salvation, however, after a period of atonement everyone will again commune with God and fully comprehend the unity of his system. In the second volume of *Observations*, Hartley thus encourages readers to cultivate imagination, sympathy, theopathy, and moral sense and provides suggestions on topics as varied as diet, sexual relationships, and aesthetic pursuits that offer guidance on how to 'make Duty itself a Pleasure, and convert Sin into a Pain' and eventually lead to more exalted states of being.[138]

However, at the end of the first volume, Hartley admits that one consequence 'flowing from the Doctrine of Association ... is thought by many to have a pernicious Tendency in respect of Morality and Religion', that is, the doctrine of necessity.[139] Most religious thinkers, including King, Gay, and Butler, believed that freewill was integral to virtue and vice and that God made humans free so that they could choose to embrace or reject him. King goes so far as to consider moral 'evil as a necessary effect of freedom'.[140] In contrast, Hartley recognizes that his theory of cognition entails necessitarianism. He states that 'each Action results from the previous Circumstances of Body and Mind in the same manner, and with the same Certainty, as other Effects do from their mechanical Causes'.[141] In other words, when faced with the same set of circumstances, people will be unable to respond differently because of association. The doctrine of necessity seems to contradict Hartley's belief in universal salvation and raises questions about why he would give suggestions for cultivating the higher intellectual affections to anyone other than those responsible for educating children. If people are doomed to repeat past mistakes, then it seems that they can never achieve salvation because they cannot help but respond to temptations exactly as they have in the past. Additionally, it would seem that people cannot change once they have fully matured because their actions are dictated by past experience.

Why then does Hartley assert 'that we have the Power of generating these Affections in ourselves, by introducing the proper generating Causes, and making the proper Associations'?[142] Most scholars have misinterpreted the doctrine of necessity, and discussions of it involve 'such oversimplifications [that they] do Hartley an injustice, since they omit all mention of Hartley's important account of voluntary action'.[143] Hartley does not eliminate the concept of freewill altogether. Rather he distinguishes between philosophical and practical freewill, defining the former as freedom from the laws of nature,

138 Hartley, *Observations*, I, p. 497.
139 Hartley, *Observations*, I, p. 500.
140 Hartford, 'William King', p. 59.
141 Hartley, *Observations*, I, p. 500.
142 Hartley, *Observations*, II, p. 53.
143 Hatch, 'David Hartley', p. 35.

which he sees as impossible because human beings cannot transcend physical laws. However, he states that 'if Free-will be defined as the Power of doing what a Person desires or wills to do, of deliberating, suspending, choosing, *&c.* or of resisting the Motives of Sensuality, Ambition, Resentment, *&c.*' it 'is not only consistent with the Doctrine of Mechanism, but even flows from it' because God designed the human body to perform voluntary and semi-voluntary actions, such as thinking, stirring up emotions, and 'performing and suspending Actions'.[144] As Hatch explains, philosophical freewill is 'the power to *will* what [one] wills' and practical freewill is 'the power to *do* what [one] wills'.[145] The former is impossible, but the latter is necessary to cultivate the intellectual affections. Practical freewill enables people to resist temptation, to choose virtuous action, and to cultivate moral growth by seeking out practices that strengthen positive habits.

Hartley occupies a crucial spot in a lineage of British organicism. Moral sense exemplifies the precepts of Hartley's organicism because it is 'both the summation of all pleasures which preceded it, and also … something more than the mere summation; it is embodied and validated its own authority'.[146] Although his doctrine of vibrations quickly lost purchase, the structures of thought that he put in place continued to create social energy well into the Romantic period and beyond.[147] Indeed, as a transitional figure between seventeenth-century belief in providence and eighteenth-century faith in progress, Hartley's conception of the cosmos as an organic system that is always in motion was key. The 'term *organization*' was used by so-called mechanists 'in the same way as the term "organism", which only became current in French (or German) by the 1800's [*sic*]'.[148] According to Charles T. Wolfe, organization refers to self-regulation, the sympathetic relationship between organs, and other 'features which were not present in machines'.[149] By this definition, Hartley does not reduce living systems to mere machines, even though he describes the cognitive processes in mechanistic terms. His account of the relationships between the mind, body, and nature instead posits them as a set of interactive processes. Thus, the two volumes of *Observations* offer an integrated account of nature, humankind, and theology that unites them into a holistic system and attempts to heal the Cartesian schism.

Furthermore, Hartley frames aesthetics in terms of embodiment, affect, and ethics and offers an understanding of transcendence that is neither ontological

[144] Hartley, *Observations*, I, p. 501.
[145] Hatch, 'David Hartley', pp. 34–35, emphasis mine.
[146] Oberg, 'The Progress Toward the Perfection of Man', p. 123.
[147] See Stephen Greenblatt, *Shakespearean Negotiations* (University of California Press, 1988) for a discussion of social energy.
[148] Wolfe, 'Determinism/Spinozism in the Radical Enlightenment', p. 48.
[149] Wolfe, 'Determinism/Spinozism in the Radical Enlightenment', p. 48.

nor disembodied; neither does it separate humankind from the phenomenal world. His ethics, like the rest of his theory, privileges relationality. His ability to weave together morality, body, mind, and the natural world provided an important model for Romantic men of science and for poets who wanted to understand how the pieces of the physical world fit together to produce intangible ideals for society. Despite the outmoded science, Hartley's model continued to provide an example of how to think the world and humankind together into the nineteenth century.

Chapter Two

Posthuman Organicism

Erasmus Darwin

In 1770, Erasmus Darwin added the inscription *e conchis omnia*, 'everything from shells', to the coat of arms emblazoned on his carriage door.[1] Combined with the three small scallop shells that were part of his family crest, this motto encapsulates his theory of evolution, which posits that all life originated from 'a single living filament'.[2] Despite the obvious differentiation among species, Darwin claimed that this evolutionary kinship was not merely historical but that all living things share similar metabolic and cognitive processes. Advancing a combined theory of life and mind that incorporates aspects of eighteenth-century vitalism and materialist theories of cognition, he posits a 'deep continuity between life and mind', which he represents as co-implicative processes that emerge together, both phylogenetically and ontogenetically.[3] In Darwin's view, as living bodies become increasingly complex, cognitive capacity also grows more sophisticated. Although an atheist, he participates in a conversation initiated earlier in the century by religious thinkers John Gay and David Hartley, who argued that the conscience, as part of the mind, matures over time as a person moves from childhood to adulthood. Along with Joseph Priestley, whom Darwin knew, Gay and Hartley formulated their theories in response to John Locke's claim that no one is born with innate ideas. In addition to extending Locke's ideas, Darwin attempts to overcome the gap between the mind, body, and world introduced by Cartesian dualism.

[1] Desmond King-Hele, *Erasmus Darwin* (Charles Scribner's Sons, 1963), p. 89.

[2] Erasmus Darwin, *Zoonomia or, the Laws of Organic Life*, 1st edn, Cambridge Library Collection, 2 vols (Cambridge University Press, 2009), i, p. 498. Unless otherwise specified, I quote from this edition.

[3] Evan Thompson, *Mind in Life: Biology, Phenomenology, and the Science of Mind* (Harvard University Press, 2007), p. ix.

He believed, like French thinker Denis Diderot, that the planet evolved over a period of millions of years. In Darwin's account, the environment and its living creatures evolved in a mutually reciprocal process, producing and shaping each other. In a relationship 'of mutual codependency; [in which] neither one nor the other is privileged', both the animate and inanimate parts of the Earth form a mutually constitutive, interdependent system.[4] Darwin's view of the relationships between different parts of the physical world offers, I argue, another British model of embodied organicism.

As a pioneer of the biological sciences, Darwin saw life and mind as a set of physical processes that occur within a larger cosmic cycle. His historicized account of life embeds living beings into a physical environment. Embodied and embedded, the mind is inseparable from the body and the world. Despite the material and physiological foundation of Darwin's system, like Hartley, he imagines the mind–body–world relationship as an organic unity that mirrors the relationality that structures the larger world. Yet his organicism differs significantly from Hartley's model as well as from German models put forward around the same time. Both Hartley and neo-Kantian thinkers grappled with the problem of how to integrate mind into nature. While Hartley combines anthropological and natural history approaches in his account of cognition, he is primarily concerned with human beings. Darwin, on the other hand, includes simple organisms such as insects and plants in his theory. Hence, he advances a concept of organicism that aligns with twenty-first-century posthumanist views of humankind's relationship to the rest of the material world. Based on his theories of life and mind, elements which he claims are driven by the same metabolic processes, Darwin's theory of evolution and his organicism level the sharp distinctions between humankind and the rest of the living world. He integrates life and mind into an organic economy that consists of living bodies as well as non-living phenomena. Thus, he sees the environment as crucial to cognition and to his concept of organicism.

Pushing the concept of organicism even further, Darwin positions organic and nonorganic entities in mutually co-dependent and co-productive relationships. The material that they comprise and the processes that enable them to interact unify these disparate parts of the world in a way that reflects the posthumanist spirit of Donna Haraway's 'Cyborg Manifesto' (1985). Darwin's organicism holds together 'incompatible things', seeing all of them as 'necessary and true'.[5] He brings the human and nonhuman worlds, both vital and

[4] Carl Knappett, 'The Affordances of Things: A Post-Gibsonian Perspective on the Relationality of Mind and Matter', in *Rethinking Materiality: The Engagement of Mind with the Material World*, ed. by Elizabeth DeMarrais, Chris Gosden, and Colin Renfrew, McDonald Institute Monographs (McDonald Institute for Archaeological Research, 2004), pp. 43–51 (p. 43).

[5] Donna J. Haraway, 'A Cyborg Manifesto: Science, Technology, and Socialist-Feminism

nonorganic, into conceptual unity by valuing them equally. Darwin also strives for unity by knocking humankind off its humanist perch and integrating people into a world made up of both objects and subjects. His theories of life and mind, along with his account of emotion, put humans on a par with other lifeforms. Where twenty-first-century new materialist Jane Bennett advances the notion of 'thing-power' to call attention to the wider world in which humankind is situated, to challenge the widespread view that it is the primary (only) actor in a passive environment, and 'to give voice to a vitality intrinsic to materiality', Darwin transforms nonhuman beings—plants and animals, objects under the Cartesian paradigm—into thinking and feeling subjects by basing subjectivity on the capacity to feel, which all creatures share to some degree. He posits one model of development and one sentient principle for all species, and representing feeling plants and the 'affection' that 'mothers of the bestial world' have for their offspring.[6] Below, I argue that Darwin's theory of character formation, which relies on the association of feeling, applies equally to animals as to human beings, thus presenting the former as feeling subjects. In contrast to but with a similar motive as Bennett, who imagines human beings as objects and highlights their thingness, Darwin conceives of all living beings as subjects and highlights their ability to be affected.

Like Hartley, Darwin includes emotion in his definition of cognition in contrast to ratiocination and reasoned deduction only. By placing emotion at the centre of his theory of embodied cognition, he underscores the similarities among living creatures and uses feeling to undermine the great chain of being. His representation of plants and animals challenges the Cartesian view of them as unthinking, unfeeling brutes in order to generate a sense of fellow feeling among his readers. Along with disseminating scientific knowledge to audiences who may not otherwise have had access to it, a project shared by his fellow French Enlightenment physician Julien Offray de La Mettrie, he encouraged readers to look at the nonhuman world more sympathetically.[7] Emphasizing the universal and embodied character of life and mind, advancing a holistic concept of organicism that includes animate and inanimate nature, and insisting on ontological unity, Darwin advocates an ethics of relationality based on feeling. In his poetry and prose, 'language, tool use, social behavior, mental events—nothing really convincingly settles the separation of human

in the Late Twentieth Century', in *Manifestly Haraway*, ed. by Cary Wolfe (University of Minnesota Press, 2016), p. 5, doi:10.5749/minnesota/9780816650477.001.0001.

[6] Jane Bennett, *Vibrant Matter: A Political Ecology of Things* (Duke University Press, 2010), p. 3; Erasmus Darwin, *The Temple of Nature: Or, The Origin of Society: A Poem, with Philosophical Notes* (J. Johnson, 1803), p. 173.

[7] Kathleen Wellman, 'Medicine as a Key to Defining Enlightenment Issues: The Case of Julien Offray de La Mettrie', *Studies in Eighteenth-Century Culture*, 17 (1987), pp. 75–89 (pp. 83–84).

and animal'.[8] In contrast to Diderot, who sees morality as unique to the human species, he deliberately unsettles the human/animal binary as he calls his readers to see themselves as part of a community that includes all species.[9]

Darwin's account of emotion underwrites his ideas about human and nonhuman society. In his view, society and language evolved from embodied emotions, the need to work cooperatively, and the desire to communicate. Extending ideas advanced by moral philosophers such as Francis Hutcheson, who placed feeling at the centre of human morality, he challenges Enlightenment binaries that privilege reason over emotion.[10] In Darwin's account of cognition, feeling develops prior to reason in human beings and animals. He also synthesizes aspects of George Cheyne's nerve theory and Edmund Burke's physiological sublime. Like Cheyne in *The English Malady* (1733), which helped initiate the cult of sensibility with the theory that sympathy originates in the nervous system, Darwin locates the origin of feeling in the sensate body.[11] In the first volume of *Zoonomia* (1794), he argues that emotion emerges from structured interactions between the body and its environment, emphasizing emotions that connect sympathy and society with aesthetics. Darwin asserts that as infants interact with their human and nonhuman environments they develop the full range of feelings, and those that are generally considered positive he associates with the aesthetic faculty. Grounding aesthetic responses in feelings of familial love and a sense of community, Darwin flips Burke's claim in *A Philosophical Enquiry into Our Ideas of the Sublime and the Beautiful* (1757, 1759) that social instincts drive the feelings that emerge from encounters with the beautiful, arguing instead that aesthetics drive sociality.[12]

In addition to the ontogenetic and phylogenetic role of aesthetics, Darwin's work highlights, according to Allison Dushane, 'the crucial role that aesthetic form had come to play in the articulation and dissemination of ideas about the increasingly complex relationship between human beings and material

[8] Haraway, 'A Cyborg Manifesto', p. 10.

[9] Lester B. Crocker, *Diderot's Chaotic Order: Approach to Synthesis* (Princeton University Press, 1974), p. 41.

[10] See Francis Hutcheson, *An Essay on the Nature and Conduct of the Passions and Affections* (J. Darby and T. Browne, 1728), Eighteenth Century Collections Online <https://link.gale.com/apps/doc/CW0119816864/ECCO?sid=bookmark-ECCO&xid=54b503ee&pg=1> [accessed 22 April 2024].

[11] See G.J. Barker-Benfield, *The Culture of Sensibility: Sex and Society in Eighteenth-Century Britain* (University of Chicago Press, 1992) for a discussion of sensibility and nerve theory. See Thomas Dixon, *From Passions to Emotions: The Creation of a Secular Psychological Category* (Cambridge University Press, 2003) for an explanation of the complicated relationship between the passions, appetites, reason, the body, and the soul prior to and during the Enlightenment.

[12] Hartley makes the same connection in *Observations on Man* (1749), which was published eight years before Burke's *Enquiry*. See chapter one.

nature'.[13] Dushane's claim is borne out by the sprawling structure of Darwin's poetry, which integrates verse, prose notes, and endnotes. Meaning in these poems emerges from their disparate parts, thus enacting the idea of organic unity, for without the discursive notes their content would be nigh impossible to comprehend, while without the verse it would cease to be poetry. However, I contend that Darwin's interest in aesthetics extends beyond form and that his theory of emotion indicates that he considers embodied aesthetic experiences as constitutive of human and nonhuman subjects and the societies in which they live. Where Hartley sees aesthetics as necessary to the greater social good because it orients people towards community, fosters social feeling, increases the propensity to behave more sympathetically, and brings them closer to God, Darwin sees it as the very foundation of society.

In Darwin's local intellectual community, Thomas Seward, a fellow member of the Lichfield literary circle, and other contemporaries understood the radical implications of his theories. Seward's satirical poem, published in *The Gentleman's Magazine* (1784), before Darwin committed his view to print, correctly alleges that he 'renounces his Creator, / And forms all sense from senseless matter; / Makes men start up from dead fish-bones'.[14] Seward's response exemplifies the conservative reaction to Darwin's work; he and others like him were clearly threatened by a theory of nature that decentred God. However, Darwin does more than simply dismiss an anthropomorphic deity. In *The Loves of the Plants* (1789), *The Economy of Vegetation* (1791), *Zoonomia*, and *The Temple of Nature* (1803), he displaces humankind. The way he situates humans in relation to animate and inanimate nature imparts his work with a potent posthuman paradigm for conceptualizing humanity's place in the cosmos. In contrast to Alan Bewell's view that Darwin celebrates human invention in the guise of tributes to nature, I argue that he advances a posthuman view that recognizes that 'the human is always intermeshed with the more-than-human-world'.[15] Bewell sees Darwin as a technocrat and global imperialist whose poetry 'enfranchise[d] nature in order to increase its value and exchangeability'.[16] Certainly, Darwin promoted capitalist ideologies, and even though he failed to address the destructive outcomes of industrialization, of which he was presumably aware, his concept of organicism conceptualizes

[13] Allison Dushane, '"Mere Matter:" Causality, Subjectivity and Aesthetic Form in Erasmus Darwin', *Romanticism and Victorianism on the Net*, no. 56 (2009), para. 22, doi:10.7202/1001100ar.

[14] Thomas Seward, 'Untitled', *The Gentleman's Magazine*, 54.2 (1784), p. 87 (ll. 23–25).

[15] Stacy Alaimo, 'Trans-Corporeal Feminisms and the Ethical Space of Nature', in *Material Feminisms*, ed. by Susan Hekman and Stacy Alaimo (Indiana University Press, 2008), pp. 237–264 (p. 238).

[16] Alan Bewell, 'Erasmus Darwin's Cosmopolitan Nature', *ELH*, 76.1 (2009), pp. 19–48 (p. 31).

relationality in important ways that are especially relevant today. Where Cartesianism conceptually separates humankind from the world that it inhabits, Darwin's vision of integration and relationship offers a more productive way to understand and orient ourselves to the human, nonhuman, and non-living world by seeing ourselves as part of rather than separate from the connections that bind us together.

In this chapter, I explore Darwin's theory of life and cognition as a response to the Cartesian crisis that culminated in the Romantic period. I argue that his twinned theory underwrites his concept of organicism, which integrates the organic and nonorganic worlds and creates a unified community of living organisms based on social feeling. I examine Darwin's ideas in the context of British and French Enlightenment critiques of Cartesianism, in regard to Hartley's and Priestley's influential theories of mind and matter, and with respect to the integrative materialist theories advanced by the likes of Diderot and La Mettrie. Additionally, I examine Darwin's ideas about society, sympathy, and aesthetics in light of eighteenth-century aesthetic philosophy. In keeping with Devin Griffith's claim that 'Darwin meant for his body of work to be read as a unified system' that 'resolves into a coherent argument', I interrogate his vision of nature, mind, and society across the body of his work.[17] *Zoonomia*, a nosology that attempts to sketch a theory of health and disease along the lines of William Cullen's system, cohesively delineates his theory of life and mind. His long poems, *Loves* and *Economy*, which together comprise *The Botanic Garden* (1789–91), and *The Temple of Nature* work together not only to educate readers about the latest scientific theories but also to advance a comprehensive system of nature and society. *Loves* personifies plants and reinforces Darwin's argument, elaborated more systematically in *Zoonomia*, that nonhuman lifeforms possess the full range of emotions and are, therefore, subjective beings. Organized into four cantos, one for each element, *Economy* positions science and technology as an extension of natural forces, while *Temple* traces the origin and evolution of life, mind, and society from inchoate matter through an endless cycle of natural processes. The poetry's fanciful figures illustrate Darwin's theories and reveal how insistently he decentres human beings by embedding them in nature. Despite some internal contradictions, taken as a whole his work presents a cohesive theory of mind and body that situates humankind within the natural order and urges us to see ourselves that way.

[17] Devin Griffiths, *The Age of Analogy: Science and Literature Between the Darwins* (Johns Hopkins University Press, 2016), pp. 79–80.

The Mental Life of Life

Darwin's organicist view of the cosmos and his commitment to medicine drove his investment in the mind–matter and vitalism debates. Building on earlier work by Hartley and Priestley, Darwin pursued their Enlightenment challenge to Cartesianism, serving as a bridge into the Romantic era. Relegating the body to the material realm, René Descartes deemed it an object bound by the laws of mechanical motion, disconnected from the active principle that impels cognition. This view authorized a host of iatromechanical theories of physiology that conceptualized the body as a machine. 'In this mechanistic neurophysiology', explains C.U.M. Smith, 'animal spirits rising from the "dark fire" in the heart percolated through the tubular fabric of the brain to fill the cerebral ventricles', and the 'valves opening in the walls of ventricles allowed these gaseous spirits to flow down the tubular nerves to activate the muscles'.[18] Isaac Newton's physics, which perpetuated Cartesian dualism by positing that matter needs an immaterial force to spur it into action, exacerbated iatrophysical approaches to the body. Thus, physicians understandably understood the body in mechanistic terms.[19] These theories spread across Europe through educators such as Herman Boerhaave under whom La Mettrie and Cullen studied, but this next generation of thinkers found the mechanical approach unsatisfying and adopted the tenets of irritability and sensibility advanced by Albrecht von Haller (also Boerhaave's student) to explain animal motion.

Having studied at the medical school in Edinburgh, Darwin attended Cullen's lectures and adopted some of his views, but he objected to iatromechanical physiology because it 'considered the body as an hydraulic machine, and the fluids as passing through a series of chemical changes, forgetting that animation was its essential characteristic' and 'sought to explain the laws of life by those of mechanism and chemistry'.[20] Darwin further condemned Cartesian-influenced theories of the body because, in his view, they led to the practice of bad medicine, causing 'unceasing injury' and 'the destruction of thousands'.[21] Additionally, these approaches sever the human subject from the natural order, cause people to view and treat other lifeforms as unfeeling machines, and denigrate the physical world as a whole, which he found objectionable. In response, Darwin expands Haller's concepts into a theory of

[18] C.U.M. Smith, 'Julien Offray de La Mettrie (1709–1751)', *Journal of the History of the Neurosciences*, 11.2 (2002), pp. 110–124 (p. 116), doi:10.1076/jhin.11.2.110.15188.

[19] John W. Yolton, *Thinking Matter: Materialism in Eighteenth-Century Britain* (University of Minnesota Press, 1983), p. 14. Yolton provides an account of the mind–matter debates in the late seventeenth and eighteenth centuries.

[20] Darwin, *Zoonomia*, I, p. 1.

[21] Darwin, *Zoonomia*, I, p. 2.

animal motion that lays the theoretical foundations of biology, which would eventually replace mechanistic approaches to life.

Positing that one process governs both life and mind, Darwin unfolds his theory of cognition in the context of both the mind–matter and the vitalism debates. His contribution to vitalist discourse can be distinguished from those of other British thinkers and from German *Naturphilosophen* in its practical focus on health, like La Mettrie's efforts in France, and in its historization of life and, by extension, mind. Catherine Packham traces the first investigations into the lifeforce to the Edinburgh Medical School in the 1740s, where Cullen and his colleague Robert Whytt sought alternatives to the mechanical models they learned as students. While vitalism often sparked politically charged debates that continued into the nineteenth century, in its simplest definition, it is the 'theory that life is generated and sustained through some non-mechanical force or power specific to and located in living bodies'.[22] Vitalists could be dualists, as were Cullen and Whytt, or materialists, like Darwin. Although both Cullen and Whytt held that all animal motion originates in the nervous system, they thought that it 'depended on the presence of a soul which acted upon the body'.[23] Cullen, however, saw motion as the joint production of the human body and the 'determining power of the environment'.[24] Adopting a model similar to Cullen's interactive organism, Darwin developed his own version of vitalism, positing the concept of sensorial power to account for 'the inscrutable self-organizing power present in organisms' that operates materially but not mechanically.[25]

[22] Catherine Packham, *Eighteenth-Century Vitalism: Bodies, Culture, Politics*, Palgrave Studies in the Enlightenment, Romanticism, and Cultures of Print (Palgrave Macmillan, 2012), p. 1. The debates between John Abernethy and William Lawrence at the Royal College of Surgeons from 1814 to 1816 centred on questions of power, both political and theological. The two men disagreed about whether the life principle is immaterial, that is, superadded to the blood, as comparative anatomist John Hunter had claimed, or a material phenomenon dependent upon physiological organization, as John Thelwall had argued twenty years earlier. Sharon Ruston contends that Abernethy's conception of the 'principle of life was akin to those symbols of oppression designed to keep people on a virtuous path' because it operated from without, not unlike government and the Church. Sharon Ruston, '"Natural Enemies in Science, as Well as in Politics": Romanticism and Scientific Conflict', *Romanticism*, 11.1 (2005), pp. 70–83 (p. 75). Associated with the French Revolution and its British supporters, the materialist position was politicized and vilified.

[23] Julius Roca, 'William Cullen (1710–1790) and Robert Whytt (1714–1766) on the Nervous System', in *Brain, Mind and Medicine: Essays in Eighteenth-Century Neuroscience*, ed. by Harry Whitaker, C.U.M. Smith, and Stanley Finger (Springer, 2007), pp. 85–98 (p. 88).

[24] Roca, 'William Cullen and Robert Whytt', p. 96.

[25] Bennett, *Vibrant Matter*, p. 65. Roca sees Cullen's model as a 'reactive organism', but his explanation makes it evident that the human body does not simply react; rather it interacts

Presenting his system as a 'theory founded on nature, that should bind together the scattered facts of medical knowledge, and converge into one point of view', Darwin positions his vitalist theory as a corrective to mechanistic medical practice, hypothesizing how the body works to formulate a catalogue of diseases and their cures.[26] In *Zoonomia*, which details his theory of organic life and how it functions in living bodies, Darwin combines John Brown's theory of excitability with Cullen's more sophisticated account of sensibility to posit sensorial power as the force responsible for life. A former student of Cullen, Brown developed and taught a simplified version of his professor's teaching, asserting that all disease results either from an excess or deficit of excitability. German physicians equated excitability with the lifeforce, which they thought mediates the 'response of organisms to the environment ... by an intrinsic activity of the organism'.[27] Brown's principle provides a model for sensorial power as a material substance 'stored in the medullary portion of the nerves and muscles' that 'corresponds to the capacity to use stimuli or, in Brown's terms, the "exciting powers" in the body', thus excitability materializes the lifeforce instead of attributing it to an immaterial force, as Cullen does, albeit indirectly.[28] In rejecting empirical methods that use 'clinical and pathological observations', Brown perhaps gave Darwin licence to speculate, of which Cullen, ever cautious about making unsubstantiated claims, disapproved.[29] For Darwin, sensorial power functions similarly to excitability, except that he puts it into a mutually co-dependent relationship with the body. An organic substance that is 'perpetually renewed by the secretion and production of it in the brain and spinal marrow', it relies as much on bodily health as the body relies on it for its wellbeing.[30] Even though the body constantly produces

with its environment. According to Denise Gigante, in the 1790s life—in contrast to forces such as electricity or magnetism—came to be understood as a vital power immanent in certain bodies that defies quantification. However, she focuses primarily on German rather than British thought. See Denise Gigante, *Life: Organic Form and Romanticism* (Yale University Press, 2009).

[26] Darwin, *Zoonomia*, I, p. 2.

[27] Nelly Tsouyopoulos, 'The Influence of John Brown's Ideas in Germany', *Medical History*, 32.S8 (1988), pp. 63–74 (p. 71), doi:10.1017/S0025727300070800.

[28] Neil Vickers, 'Coleridge, Thomas Beddoes and Brunonian Medicine', *European Romantic Review*, 8.1 (1997), pp. 47–94 (p. 65). Brown developed his system in *Elementa Medicinae* (1780). For a fuller explanation of Brunonianism, see Vickers, 'Coleridge and Brunonian Medicine', pp. 65–68; Christopher Lawrence, 'Cullen, Brown and the Poverty of Essentialism', *Medical History*, 32.S8 (1988), pp. 1–21, doi:10.1017/S0025727300070770; Tsouyopoulos, 'The Influence of John Brown'; Guenter B. Risse, '"Doctor William Cullen, Physician, Edinburgh": A Consultation Practice in the Eighteenth Century', *Bulletin of the History of Medicine*, 48.3 (1974), pp. 338–351.

[29] Risse, 'William Cullen', p. 2.

[30] Darwin, *Zoonomia*, I, p. 75.

and expends sensorial power, if in illness it is diminished 'weakness ensues'.[31] If too much is lost, the organism will die. Conversely, if the body becomes too weak during a long-term illness or as part of the ageing process, it loses its ability to produce enough sensorial power to sustain itself.

The relationship that Darwin establishes between the body and sensorial power evinces a deep connection between life and mind. In addition to animal motion, that is, life, he attributes all cognitive activity to sensorial power, making the same force responsible for all life functions, from muscular motion to mental action. While sensorial power helped him to solve the Cartesian problem of how '*two things can influence or affect each other, which have not some property in common to both of them*' by positing only one substance, it raised another obstacle, especially for contemporaries who held the Cartesian or Newtonian theory that matter is inert and extended.[32] French philosophers, such as La Mettrie, Diderot, and Baron d'Holbach—all materialists—asserted that matter and motion, far from being separate principles, are 'essentially dependent on each other'.[33] Diderot attributes motion to 'the molecule endowed with the quality proper to its own nature, [which] is in itself an active force'.[34] Less concerned with matter's motility than the 'Penetrability of Matter', or the idea that 'two bodies can exist together in the same part of space', Darwin draws on Priestley's and Roger Joseph Boscovich's theories of matter.[35]

Priestley combines Isaac Newton's notion of action-at-a-distance with Boscovich's *A Theory of Natural Philosophy* (1758) to challenge the Cartesian characterization of matter as inert, solid, and impenetrable.[36] He theorizes that 'objects extended in space were to be explained as clusters of dimensionless points subject to oscillatory curves'.[37] According to Priestley:

[31] Darwin, *Zoonomia*, I, p. 75.

[32] Darwin, *Zoonomia*, I, p. 115.

[33] Julien Offray de La Mettrie, *La Mettrie: Machine Man and Other Writings*, trans. and ed. by Ann Thomson, Cambridge Texts in the History of Philosophy (Cambridge University Press, 1996), p. 47.

[34] Denis Diderot, *Diderot, Interpreter of Nature: Selected Writings*, ed. by Jonathan Kemp, trans. by Jean Stewart and Jonathan Kemp, 2nd edn (International Publishers, 1963), p. 128.

[35] Darwin, *Zoonomia*, I, p. 113.

[36] Alan Beretta explains the revolutionary implications of Newton's 'proposed action-at-a-distance, or gravity, to account for planetary motion', which, he claims, continues to be misunderstood today. Alan Beretta, 'Joseph Priestley: An Instructive Eighteenth Century Perspective on the Mind-Body Problem', in *Brain, Mind and Consciousness in the History of Neuroscience*, ed. by C.U.M. Smith and Harry A. Whitaker, History, Philosophy and Theory of the Life Sciences, 6 (Springer, 2007), pp. 75–89 (p. 77). He argues that in the eighteenth century Priestley alone understood the import of Newtonian physics, but he does not take Hartley's or Darwin's work into account.

[37] Karis Muller, 'Point Atomism, Space and God (1760–80)', *The European Legacy*, 12.3 (2007), pp. 277–292 (p. 218).

> principles of the Newtonian philosophy were no sooner known, than it was seen how few, in comparison, of the phenomena of nature, were owing to solid matter, and how much to powers, which were only supposed to accompany and surround the solid parts of matter. It has been asserted, and the assertion has never been disproved, that for any thing we know to the contrary, all the solid matter in the solar system might be contained within a nut-shell, there is so great a proportion of the void of space within the substance of most solid bodies.[38]

Priestley offers an architectonic model of matter that asserts that all physical bodies are made up of relatively few atoms held in place by the forces of attraction and repulsion, atoms which are the Newtonian version of Diderot's active molecule.[39] If objects lack solidity because they are composed of tiny points of matter with space in between, then materiality does not preclude the activities that dualists ascribe to an immaterial substance, including the ability to initiate movement and to cogitate. Priestley's theory of matter thus obviates the need to explain neural transduction by mechanistic means, subtle aethers, or vibrations.

Just as Priestley rejected vibrations in his 1775 abridged version of Hartley's *Observations on Man*, so too did Darwin, hypothesizing instead that sensorial power is responsible for transmitting sensory information through organs of sense and along the nervous system. He compares it to other material forces such as gravity, which 'exist without possessing the properties, which excite our smell or taste', to explain how it can be both material and imperceptible.[40] Darwin claims that sensorial power not only impels motion but also converts physical stimuli into a form that can be processed by the body and transmitted to the mind. In contrast to Brown, who attributes movement to a single substance—excitability—he adapts Cullen's taxonomy of disease, dividing its movements into four types: 'irritability, sensibility, voluntarity, and associability', which are also the four types of disorders in his nosology.[41] Following Haller, Darwin defines irritability, which is akin to sensation, as the body's capacity to sense stimuli in the external world and its involuntary muscular responses such as twitching and blinking. Sensibility is the body's capacity

[38] Joseph Priestley, *Disquisitions Relating to Matter and Spirit* (Kessinger Publishing, 2005), p. 17.

[39] See Robert Schofield, 'Joseph Priestley on Sensation and Perception', in *Studies in Perception: Interrelations in the History of Philosophy and Science*, ed. by Peter K. Machamer and Robert G. Turnbull (Ohio State University Press, 1978), pp. 336–354, for a discussion of Priestley's theory of matter. See Muller, 'Point Atomism, Space and God', for a comparison of Priestley's and Boscovich's theories.

[40] Darwin, *Zoonomia*, I, p. 114.

[41] Darwin, *Zoonomia*, I, p. 32.

to perceive sensory stimuli by transmitting it from the organs of sense to the nervous system, which is the first step in the cognitive process. Next is voluntarity, which is the ability to choose how to respond to percepts. According to Darwin, this ability to deliberate and consider how to solve problems is a cognitive act, no matter how rudimentary. Finally, associability denotes the body's propensity to develop habituated responses to repeated stimuli.

To account for sensorial power's cognitive function, that is, its ability to transform sensation into percepts—the initial stage in forming ideas—Darwin asserts that it assumes the properties of whatever objects it encounters. For example, 'at the time it receives motion from light, it must possess that property, which light possesses, to communicate that kind of motion.... And at the time it is stimulated into other kinds of animal motions by the particles of sapid and odorous bodies affecting the senses of taste and smell, it must resemble these particles of flavour and odour'.[42] Sensorial power can assume different characteristics because it shares the same basic substrate as other physical phenomena. It is thus a material phenomenon, presumably a form of energy, that occupies the same space as the body and takes on sensory data from the things it encounters.[43] Claiming that sensorial power can 'occupy space without detruding other bodies from it', Darwin accounts for both its embodiment and its embeddedness, decoupling materiality from solidity, and thus obviating the need to resort to dualism.[44] This theory of matter plays a significant role not only in Darwin's theory of life but also in his account of evolution.

Darwin's model of evolution locates life temporally. He saw the history of the Earth, as did Diderot and Holbach, in terms of millions rather than the thousands of years of the Bible. However, in contrast to Diderot, who did not have a 'theory of organic evolution, with its tight interrelationship' between organisms and environments, Darwin's ideas about the creation of the universe not only illustrate the reciprocal relationship that he envisioned between life and

[42] Darwin, *Zoonomia*, I, p. 115. As fantastic as Darwin's ideas may sound, they are not entirely inconsistent with current scientific theory. The wave-particle duality of light provides a useful analogy for thinking about his claim. Photons, 'the fundamental particles of light', which 'have a unique property in that they are both particle and wave' are necessary for the refraction and diffusion of light. Tega Jessa, 'What Are Photons', *Universe Today*, 19 September 2010 <https://www.universetoday.com/74027/what-are-photons/> [accessed 23 July 2019]. Light, like sensorial power, sometimes has one kind of characteristic and at other times a completely different one. Thanks to Benjamin Hagen for bringing this bit of science to my attention.

[43] Martin Priestman notes that the concept of energy is crucial to Darwin's materiality, 'without [which] matter becomes a passive machinery, requiring some force to animate it: the "ghost in the machine" of Cartesian dualism'. Martin Priestman, *The Poetry of Erasmus Darwin: Enlightened Spaces, Romantic Times* (Ashgate, 2013), p. 103.

[44] Darwin, *Zoonomia*, I, p. 114.

mind, but they also reveal the extent to which he saw these twin phenomena as embodied and embedded in the physical world.[45] Before life and mind can obtain, in his view, they require a substratum that can support them, both in the present and historically. Examined through the lens of the twentieth-century biological theory autopoiesis, it is clear that Darwin saw life not only as co-implicative with mind but also as self-organizing. Life and mind thus emerge together in his theory because they are both self-organizing. Formulated by neurobiologists Humberto Maturana and Francisco Varela, the autopoietic account of life distinguishes between non-living and living systems based on how they come into being. Heteronomous processes that exert pressure from the outside shape non-living systems. In eighteenth-century terms, they are determined by the mechanical forces associated with Cartesian and Newtonian physics. Living systems, however, organize themselves autonomously from within. In conjunction with the principles of dynamic systems theory, which explain how complex systems made up of a collection of components become something more than the sum of their parts, autopoiesis helps to account for organicism from a biological perspective. In his account of the origin of life, Darwin situates animate and inanimate nature in a mutually co-dependent relationship. Even though nonorganic matter evolved first, he does not make it ontologically prior. In his view, sensorial power existed in a latent state before life emerged. Diderot's claim that 'in this universe everything is either in translation or *in nisu*, or else in translation and *in nisu*' clarifies this detail of Darwin's theory.[46] All matter is either in an active state of structuration and movement or it is in a state of potential, and sensorial power exists in the latter state even when it does not have a substrate to support it, like electricity in the absence of suitable conductors.

While Darwin discusses evolution in his chapter on reproduction in *Zoonomia*, his representation of it in *Temple of Nature* and *Economy of Vegetation* offers more insight into the details, including how he sees the relationships between the non-living and living domains. The interdependence among the various parts of nature reflects Darwin's concept of organic unity. Martin Priestman notes that his poetry emulates Lucretius's didactic epic, which also advances an explanation of the natural world in terms of physical laws rather than divine intervention.[47] Formally, Darwin's perfectly metred and rhymed poetic couplets exemplify the classical aesthetic associated with Alexander Pope and the earlier part of the eighteenth century, but in many ways *Temple*

[45] Crocker, *Diderot's Chaotic Order*, p. 23.

[46] Diderot, *Interpreter of Nature*, p. 127.

[47] Priestman, *The Poetry of Erasmus Darwin*, p. 20. See Amanda Jo Goldstein, 'Nerve Poetry and Fiber Art: Biosemiosis and Plasticity in Erasmus Darwin', *Literature Compass*, 17.3–4 (2020), 17:e12552, doi:10.1111/lic3.12552 for an analysis of Lucretian materialism in the Romantic era.

is also a secular version of John Milton's *Paradise Lost* (1667), particularly Books VII and VIII in which 'Raphael at the request of Adam relates how and wherefore this world was first created'.[48] However, Darwin replaces the Christian framework with the Rosicrucian machinery of gnomes, sylphs, and the like, and, of course, he challenges Cartesian ontology.[49] Yet *Temple* 'uses much larger-scale ideas of time to re-shape the vision' of nature and society in terms of the 'evolution of the earth and its living species'.[50] In this version of the creation story, the garden does not come first, but rather the environment and its inhabitants develop together. In contrast to his grandson's account of evolution, which privileges adaptability to environmental conditions, in Darwin's account organisms and their ecological niches evolve together dynamically. However, Darwin makes it clear that while Newtonian physics govern the evolution of the non-living parts of the globe, the laws of sensorial power govern life.

Considering Darwin's representation of creation from an autopoietic perspective, the inanimate environment forms heteronomously in contrast to the autonomously organizing processes of living beings. *Economy* provides an account of how the universe began with a big bang, albeit in nominally Christian terms:

> —LET THERE BE LIGHT!" proclaim'd the ALMIGHTY LORD,
> Astonish'd Chaos heard the potent word;
> Through all his realms the kindling Ether runs,
> And the mass starts into a million suns;
> Earths round each sun with quick explosions burst,
> And second planets issue from the first.[51]

48 John Milton, *Paradise Lost*, ed. by Alastair Fowler, Longman Annotated English Poets, 2nd edn (Longman, 1998), p. 388.

49 Although Darwin occasionally refers to 'GOD THE FIRST CAUSE!' or even 'the great First Cause', he was in all probability an atheist. Darwin, *Temple*, 1.3.233; Darwin, *Zoonomia*, I, p. 505. According to his most prolific biographer, Desmond King-Hele, he had a 'habit of scoffing at Christianity', a report corroborated by Anna Seward's *A Memoir of the Life of Dr Darwin* (1804). King-Hele, *Erasmus Darwin*, p. 23. In *Zoonomia*, Darwin claims that the origin story in Genesis is allegorical: 'This idea of the reproduction of animals from a single living filament of their fathers, appears to have been shadowed or allegorized in the curious account in the sacred writ of the formation of Eve from a rib of Adam'. Darwin, *Zoonomia*, I, p. 489. Priestman also contends that Darwin's 'loudly capitalized references to God ... are undercut' by his notes which clarify that he sees the origin of the cosmos in scientific rather than theological terms. Priestman, *The Poetry of Erasmus Darwin*, p. 107. Thanks to Dahlia Porter for bringing the Rosicrucian symbology to my attention.

50 Priestman, *The Poetry of Erasmus Darwin*, p. 103.

51 Erasmus Darwin, *The Economy of Vegetation. A Poem. With Philosophical Notes*, in *The*

In *Temple*, Darwin expands his account of the formation of the universe while simultaneously zooming in to focus more specifically on the Earth and its solar system. Dispensing with religious language, he attributes creation not to God but to Newtonian forces.[52] In this more condensed version, Darwin's language ascribes causative power to heat:

> First HEAT from chemic dissolution springs,
> And gives to matter its eccentric wings;
> With strong REPULSION parts the exploding mass,
> Melts into lymph, or kindles into gas.[53]

Darwin assigns active, present tense verbs to heat, which first 'springs' from the 'chemic' stew before it acts on matter, causing it to explode, melt, and kindle. Despite this attribution of agency, he is clearly not describing life, but rather 'REPULSION' and attraction, two mechanical forces. Repulsion causes the original explosion, and in the next quatrain, 'ATTRACTION' 'Approaching parts with quick embrace combines, / Swells into spheres and lengthens into lines' the atoms of matter that eventually form the Earth's terrain.[54] These verses illustrate Newtonian forces acting heteronomously on disorganized bits of matter, pushing and pulling them into architectonic forms that become the terraqueous globe which, in turn, provides a hospitable environment for life to emerge. As the forms continue to be shaped by external forces, they eventually differentiate into fibrous structures capable of supporting life by providing a substrate for sensorial power.

Describing the formation of bodily structures, Darwin makes it clear that while an external source spurs the fibres into action, an intrinsic one sparks them into life. Darwin's attribution of agency moves from external (heteronomous) to internal (autonomous) forces:

> Last, as fine goads the glutton-threads excite,

Botanic Garden: A Poem in Two Parts. Part I. Containing The Economy of Vegetation. A Poem. With Philosophical Notes, 3rd edn (J. Johnson, 1795), 1.103–108.

[52] In *Zoonomia*, Darwin makes it clear that he is not interested in debating matters of religion. He states that he does 'not wish to dispute about words, and am ready to allow, that the powers of gravity, specific attraction, electricity, magnetism, and even the spirit of animation, may consist of matter of a finer kind; and to believe, with St. Paul and Malebranch [*sic*], that the ultimate cause only of all motion is immaterial, that is God'. Darwin, *Zoonomia*, I, p. 109. Like the French materialists, who needed to practise 'rigid self-censorship to escape fatal offence to the government or the Sorbonne', Darwin hedged about religious matters to mitigate the atheistic implications of his theories. P.N. Furbank, *Diderot: A Critical Biography* (A.A. Knopf, 1992), p. 86. He shrewdly paid just enough lip service to Christianity to avoid alienating his upper- and middling-class patients.

[53] Darwin, *Temple*, 1.4.235–238.

[54] Darwin, *Temple*, 1.4.240–242.

> Cords grapple cords, and webs with webs unite;
> And quick CONTRACTION with ethereal flame
> Lights into life the fibre-woven frame.—[55]

Once touched by 'fine goads', the cords begin to actively 'grapple' with each other, while 'webs' take the initiative to 'unite' with other 'webs'. Once stimulated into activity, they contract, which ignites them into life. A footnote explains that contraction is a force that 'exists in organized bodies, and distinguishes life from inanimation' by causing nerve fibres to shorten.[56] It is the action of sensorial power in the mode of irritability. This note reaffirms that sensorial power is an autonomous force and confirms that life is not extrinsic to the body, like Hunter's and Abernethy's vital spark. Where Diderot 'suspect[s] that animality had from all eternity its particular elements scattered in and mingled with the mass of matter', Darwin attributes the origin of life to self-organization.[57] In these lines, sensorial power is the intrinsic capacity of certain kinds of matter to self-organize. Even before the fibres are stimulated to contract, Darwin's language suggests a readiness to assume agency. He describes them as 'glutton-threads', which suggests motive—hunger. Desire, according to Darwin, is the precursor to agentic action. Although the fibres are not initially alive, they are primed to respond not mechanistically but by directing the activity of contraction inwards towards themselves. 'Hence', says Darwin,

> without parent by spontaneous birth
> Rise the first specks of animated earth;
> From Nature's womb the plant or insect swims,
> And buds or breathes, with microscopic limbs.[58]

These lines more clearly depict the teeming, self-motivated action of lifeforms, as the active verbs—'Rise', 'swims', 'buds or breathes'—indicate. The next line reinforces life's immanence in the natural order with the repetition of the preposition 'in': 'In earth, sea, air, around, below, above, / Life's subtle woof in Nature's loom is wove', and the lines that follow demonstrate that it is also immanent in individual organisms.[59] As living matter shapes itself, Darwin continues to use active verbs to describe the motion of new life in autopoietic terms. In autopoiesis (a neologism that means self-making), organisms distinguish themselves and remain distinct from their environments by forming

[55] Darwin, *Temple*, 1.4.243–246.
[56] Darwin, *Temple*, n. 1.4.245.
[57] Diderot, *Interpreter of Nature*, p. 48.
[58] Darwin, *Temple*, 1.4.247–250.
[59] Darwin, *Temple*, 1.4.251–252.

boundaries that separate inside from outside. The boundary contains the structures required for the metabolic processes that keep living things alive. Darwin's description is highly suggestive of boundary formation:

> Points glued to points, a living line extends,
> Touch'd by some goad approach the bending ends;
> Rings join to rings, and irritated tubes.[60]

Darwin's image of a line that bends towards itself to form a ring is striking both because it illustrates the process of autonomous organization and because it implies boundary formation. Grammatically, 'the living line' becomes the subject that 'extends' itself into rings and tubes. This verse implies that simple organisms evolve into creatures with increasingly complex body structures.

The metaphorical boundary drawn by the 'living line' separates it from the inanimate environment in which it is embedded and maintains the organism's structural integrity by enclosing the assemblage. However, it must be semipermeable so that the organism can interact with the environment to nourish itself by imbibing food and avoid danger by sensing nearby threats, and to flourish as a result. Darwin's illustration of the newly emergent organisms suggests that this boundary is, indeed, semipermeable, thus facilitating exchange with its surround as they

> Clasp with young lips the nutrient globes or cubes;
> And urged by appetencies new select,
> Imbibe, retain, digest, secrete, eject.[61]

Darwin makes it clear that the organisms exist in productive interchange with their environment. The semipermeable boundary is a crucial component of autopoietic theory because it enables the mutual dependence of parts and whole that is essential to organicism. The semipermeable boundary is necessary because it enables the filaments to ingest nourishment and makes possible an interdependent relationship with the whole.

This interactive process is also cognitive in that it provides the 'specific condition that individuates the presence of a unified cognitive process or system'.[62] Examined through the lens of autopoietic enaction, a theory that

[60] Darwin, *Temple*, 1.4.253–255.
[61] Darwin, *Temple*, 1.4.256–258.
[62] Giulia Piredda, 'The Mark of the Cognitive and the Coupling-Constitution Fallacy: A Defense of the Extended Mind Hypothesis', *Frontiers in Psychology*, 8 (2017), pp. 1664–1078 (p. 3), doi:10.3389/fpsyg.2017.02061.

extends autopoiesis into an argument for embodied and embedded cognition, these verses suggest that Darwin's first lifeforms exhibit rudimentary cognitive activity. According to enaction, cognition occurs as organisms interact with their environment, and their physical embodiment dictates, in part, what they perceive and how they interpret it. When an organism encounters stimuli, it evaluates and acts on them relative to its twofold immanent purpose: self-production, or the kind of activities it undertakes to maintain itself as a living being, and sense-making, the process of evaluating and determining how to engage with its surroundings.[63] When an organism encounters a 'perturbation'—an environmental stimulus that it notices because of its relevance to its wellbeing—it moves towards the stimuli that facilitate its life processes and away from those that threaten them.[64] According to Varela, Evan Thompson, and Eleanor Rosch, because an organism must evaluate stimuli and determine whether they help or hinder its immanent purpose, these basic life functions entail low-level cognitive activity. As an organism encounters countless stimuli, it assesses them so that it can make sense of its surroundings. As part of this evaluative process, an organism 'performs an interpretation in the sense that it selects or brings forth a domain of significance out of the background of its random milieu'.[65] In this theoretical model, life and mind are inextricably linked because the most basic life functions require organisms to appraise, discriminate, and signify what they encounter in their world.

Darwin's representation of newly emerged organisms offers an extended illustration of sense-making. To return to the lines that depict them being 'urged by appetencies new', it is clear that they seek sustenance.[66] Once they 'Clasp' their 'lips' to the 'nutrient globes or cubes', the organisms 'select / Imbibe, retain, digest, secrete, eject'.[67] This concise representation of metabolic life processes implies the first prong of immanent purposiveness, self-maintenance, while Darwin's active verbs infer the second part, sense-making.[68] The phrase 'urged

[63] Thompson, *Mind in Life*, pp. 148–149.

[64] Varela, Thompson, and Rosch, *The Embodied Mind*, p. 152.

[65] Varela, Thompson, and Rosch, *The Embodied Mind*, p. 156.

[66] Darwin, *Temple*, 1.4.257.

[67] Darwin, *Temple*, 1.4.256, 257–58.

[68] In Amanda Jo Goldstein's analysis of this passage, she concludes, based on Darwin's use of passive and active voice, that 'it ultimately makes little sense to ascribe the fundamentally relational process of embryogenesis in Erasmus Darwin either to an active, self-organizing power internal to the organism or to its purely passive determination by its surround'. Goldstein, 'Nerve Poetry and Fiber Art'. However, Tsouyopoulos's explanation of Germanic Brunonianism provides insight into Darwin's understanding of the embryonic organism's relationship with its environment. Twelve years after the publication of *Elementa Medicinae*, German physicians Andreas Röschlaub and Adalbert Friedrich Marcus formulated their own system, '*Erregbarkeitstheorie* (excitability theory)' based on Brown's excitability. Tsouyopoulos, 'The Influence of John Brown', p. 64. In their interpretation of Brown's theory,

by appetencies' suggests that the young organisms are motivated by metabolic drives to look for nourishment. Consequently, they move towards those things in their environment that will sustain them—nutrients. The verb 'select' denotes a degree of choice, which implies that they assess and screen stimuli. Indeed, the next canto explicitly illustrates the living filament's capacity to choose. As it encounters the 'elements', it 'Rejects the noxious, and the pure digests'.[69] Here, the organism exhibits the ability to discriminate between stimuli that threaten and those that promote its wellbeing. The capacity to select, imbibe, retain, and reject suggests that even the simplest organisms decide which perturbations to attend and which to ignore. Darwin's filaments exhibit the rudimentary sense-making described by enactivists.

Darwin's theory of life and mind extends agency to even the simplest lifeforms, which implies an ability to transform an environment into a milieu, or lifeworld, and to make choices about how to interact with it, albeit crudely. The drive to ingest food resembles Baruch Spinoza's concept of conatus, which Gilles Deleuze usefully explains as 'the effort by which each thing strives to persevere in its being, each body in extension, each mind or each idea in thought'.[70] Expanding on this idea 'in current biological terms', Antonio Damasio defines conatus as the 'aggregate of dispositions laid down in brain circuitry that, once engaged by internal or environmental conditions, seeks both survival and well-being'.[71] These contemporary glosses make clear that for many theorists basic metabolic drives, far from being blind, are striving, effortful, engaged, and seeking. To 'persevere in its being' implies sense-making, as an organism must determine which parts of its environment are helpful and which are inimical to its survival.

The 'appetencies' that urge Darwin's newly formed filaments to ingest food require them to confer significance on parts of their surroundings. The filaments are analogous to bacteria in a sugar gradient that move towards higher concentrations of sugar and away from chemicals that threaten to dissolve their membranes. One could explain the bacteria's response in terms of the 'local effects of sucrose on membrane permeability, medium viscosity, hydro-mechanics', all of which emphasize external factors rather than the bacteria's

'organisms possess intrinsic activity, but this has no actual reality unless the organism is stimulated from the outside'. Tsouyopoulos, 'The Influence of John Brown', p. 67. On this view, sensorial power is *in nisu* until it is spurred into action, at which point it organizes itself from within but in response to the surround, which provides the microelements necessary for life to obtain.

[69] Darwin, *Temple*, 2.1.37–38.

[70] Gilles Deleuze, *Spinoza: Practical Philosophy*, trans. by Robert Hurley (City Lights Books, 1988), p. 21.

[71] Antonio Damasio, *Looking for Spinoza: Joy, Sorrow, and the Feeling Brain* (Harcourt, 2003), p. 36.

internal motivations.[72] However, in an enactive account of cognition, these local effects become significant only in relation to the bacteria. Sugar is not intrinsically food and can be toxic to some organisms, but to the bacteria it is meaningful as food.[73] Rather than viewing organisms' behaviour as objective facts, therefore, according to the tenets of enactive cognition, they select details that are subjectively significant to them. The significance of sugar obtains relative only to bacteria. At the same time, there are plenty of things in the environment that bacteria, or the living filaments, ignore because they have no subjective relevance. In phenomenological terms, organisms make sense of the stimuli in their surround 'according to what they are worth for [their] typical activity'.[74] In Darwinian terms, irritation, the basic ability to sense one's surround, leads to sensibility, which transmits the sense data gathered by the 'extreme parts of [the sensorium], which reside in the muscles, or organs of sense'.[75] Once the filaments take action to pursue nutrients, they have exhibited volition. These three 'MOTIONS OF THE SENSORIUM' comprise the first three cognitive processes.[76] As in the enactive account of cognition, even these basic interactions between an organism and its environment constitute cognition.

Darwin explains volition, or self-motivated action, in *Zoonomia*. He associates this ability to take considered or planned action with cognitive rather than involuntary activity. He uses another food example to illustrate the difference between sensibility and volition. Gathering apples from a tree requires 'associated motions [that are] introduced by their connection with sensation'.[77] The impulse to pick and eat apples in response to hunger requires no thought. If, however, the apples are out of reach, 'desire is produced' and one 'consequence of this desire is … deliberation'.[78] Desire drives volition because it requires the ability to decide what action to take to satisfy it. Darwin explicitly connects volition with cognition when he establishes purposeful action as a 'criterion to distinguish *voluntary acts or thoughts* from those caused by sensation'.[79] By using a food-related example that attaches desire to bodily needs, Darwin, like Gay and Hartley, associates cognition with the desire to

[72] Francisco J. Varela, 'Autopoiesis and a Biology of Intentionality', in *Autopoiesis and Perception: Proceedings of a Workshop Held at Dublin City University on August 25th & 26th*, ed. by Finbarr (Barry) V. McMullin and Noel Murphy (Dublin City University, School of Electronic Engineering Technical Report, 1992), pp. 4–14 (p. 7).

[73] Erin F. Boulanger and others, 'Sugar-Phosphate Toxicities', *Microbiology and Molecular Biology Reviews*, 85.4 (2021), pp. 1–28 (p. 3).

[74] Maurice Merleau-Ponty, *The Structure of Behavior*, trans. by Alden L. Fisher (Beacon Press, 1963), p. 130.

[75] Darwin, *Zoonomia*, I, p. 40.

[76] Darwin, *Zoonomia*, I, p. 40.

[77] Darwin, *Zoonomia*, I, p. 58.

[78] Darwin, *Zoonomia*, I, p. 59.

[79] Darwin, *Zoonomia*, I, p. 59.

avoid pain or to attain pleasure, both of which originate in sensation. Enactive philosopher Ralph D. Ellis equates this effortful striving with agency, which he claims is 'the ability of the organism to direct the use of its own parts, including and especially at the micro-level'.[80] Thus, the sense-making that Darwin attributes to the first organisms indicates a degree of self-motivated power that enables them to choose aspects of their environment to engage actively.

Darwin's emphasis on the interaction between organisms and their environments is not 'merely structural', as D.B. Ruderman claims in his discussion of Darwinian subject-formation, but represents structured engagement as a form of agency.[81] Cognition is a 'structure of engagement with the world' driven by the need to sustain life and that differs by degree depending on an organism's structure.[82] In another example, Darwin cites the actions of a plant that unfurls its leaves towards the sun, a source of sustenance, and curls its leaves inwards to protect itself from the cold, which he attributes to sensibility and volition. He insists not only that 'flowers shew evident marks of sensibility' but that plants' responses to external stimuli provide evidence of 'some degree of voluntary power'.[83] Ascribing volition to plants indicates that Darwin considers their responses to the world as a form of cognitive activity.

In the eighteenth-century context, Darwin's ideas about agency are consistent with the modest views of Priestley and Hartley. Rather than the unfettered, transcendental power to will, which Hartley calls 'philosophical Freewill', Darwinian agency is akin to Hartley's practical freewill.[84] This sort of freewill allows for the power of choice but also acknowledges that it is materially conditioned. In Darwin's system, agency originates with appetency, or immanent purposiveness, which, as in Immanuel Kant's theory, serves as an organism's means and end, but rather than being a regulative principle it is constitutive. The example of organisms gravitating towards and ingesting food demonstrates the way environmental and physiological affordances impel motive but do not wholly constrain it. In *Zoonomia*, Darwin asserts that 'to will without motive is to will without desire or aversion', an idea so absurd that he calls it a solecism.[85] Conceptually linked to appetency, desire and aversion can thus only ever be experienced in relationship. A living being wants or wants to avoid some thing. To will, then, can never be a disinterested

[80] Ralph D. Ellis, *Curious Emotions: Roots of Consciousness in Motivated Action* (John Benjamins Publishing, 2005), p. 195.

[81] D. B. Ruderman, *The Idea of Infancy in Nineteenth-Century British Poetry: Romanticism, Subjectivity, Form* (Routledge, 2016), p. 77.

[82] Thompson, *Mind in Life*, p. 78.

[83] Darwin, *Zoonomia*, I, pp. 103, 104.

[84] Hartley, *Observations*, I, p. 511.

[85] Darwin, *Zoonomia*, I, p. 134.

act but must always have an intentional target, and agency operates within a system that it cannot transcend. In *Zoonomia* and *Temple of Nature*, Darwin suggests that cognition entails agency. In *Loves of the Plants*, he suggests that it also entails subjectivity.

The Inner Life of Life

If agency is the power to affect oneself and the external world in Darwin's theory of organicism, subjectivity is the capacity to be affected. In contrast to Thomas McFarland's claim that 'the self will diminish to a mere category of thing' in philosophical systems that take the object rather than the subject as their starting point, I argue that Darwin's materialism does not diminish subjectivity but expands it by extending it to other lifeforms rather than considering it the exclusive property of human beings.[86] To consider the matter from the perspective of new materialism, in which theorists such as Bennett objectify the subject to call attention to the nonhuman world, Darwin subjectifies the object much to the same end. Embedding the agent within a material system, he locates subjectivity spatially and temporally within individual organisms. Autopoietic enaction provides a framework for understanding how Darwin's lines and tubes become subjects. Thompson explains that the semipermeable boundary—Darwin's ring shaped organism—'gives rise … to inwardness and outwardness' that not only demarcates 'an inside' but 'also specifies an outside to which it is normatively related'.[87] The boundary constitutes an individual by keeping each living thing 'numerically distinct from other things' while also providing a 'self-pole' that is 'in dynamic relationship with alterity'.[88] The boundary and everything inside it organize the subject, while everything outside it forms the other, or not-self. Considered physiologically, the interior space encloses the structures and processes that drive appetency, motive, and agency. Thus, like cognition, subjectivity is determined by structures and structured interactions. The localized body provides a physical space for lived experience to constellate, and which serves as the basis of selfhood. With respect to sense-making, the meanings that an organism attributes to 'the entities exercising motivating force', many of which are external, result from 'certain forms of intentional consciousness and intersubjective processes'.[89] Likewise, in Darwin's account of subjectivity, consciousness and intersubjective

[86] McFarland, *Coleridge and the Pantheist Tradition*, p. 57.

[87] Thompson, *Mind in Life*, p. 163.

[88] Thompson, *Mind in Life*, p. 153.

[89] Christian Beyer, 'Edmund Husserl', in *The Stanford Encyclopedia of Philosophy*, ed. by Edward N. Zalta and Uri Nodelman, Summer 2018 (Metaphysics Research Lab, Stanford University) <https://plato.stanford.edu/archives/sum2018/entries/husserl/> [accessed 5 December 2019].

experiences play an important role in its formation, but the former is not its sole ground.

Darwin's account of subject-formation in *Zoonomia* challenges Enlightenment humanism and Cartesian dualism by refusing to privilege human consciousness. Contrary to Cartesian epistemology, Darwin sees consciousness as a mode of cognition rather than as its ground. Dissecting the *cogito*, he contends that dualist ideas about consciousness are tautological, stating 'I believe it is Des Cartes, [who] has said, "I think, therefore I exist." But this is not right reasoning, because thinking is a mode of existence; and it is thence only saying, "I exist, therefore I exist"'.[90] According to Darwin, existence has three facets: being, taking action, and being acted upon.[91] Being corresponds to what Stefan Esposito calls 'bare biological life' or the mere 'physiological aspects of life'.[92] The latter two correspond to agency and subjectivity. Darwin further states that the 'when, and the where' of 'this existence, depends on the successive motions of our own or of other bodies'.[93] In contrast to an absolute consciousness that abides after death, he makes the 'radical suggestion of a contingent form of subjectivity, built on the foundation of a complex system of causality'.[94] Subjectivity depends on embodiment and embeddedness. Additionally, Darwin recognizes that people are not conscious during every waking moment. He asserts that 'We are only conscious of our experiences when we think about it; ... when we are busied about other objects, neither the lapse of time nor the consciousness of our own existence can occupy our attention'.[95] Because humans lack self-awareness much of the time, consciousness cannot be the ground of subjectivity.

Darwin's definition of consciousness is specific to humankind, setting it apart from the rest of the living world. However, neither is it the most important thing about humanity nor does it set humans above other species. Darwin defines consciousness as the awareness 'of a time of our existence previous to the present time'.[96] This awareness of 'identity now and heretofore' enables people to recollect, compare, and narrativize their sense of self.[97] As

[90] Darwin, *Zoonomia*, I, p. 133.

[91] Darwin's shorthand for these three states of being are 'I am', 'I move', and, oddly, 'I suffer'. Darwin, *Zoonomia*, I, p. 133.

[92] Stefan Esposito, 'Communicating Life: Mary Shelley's *Frankenstein* and Romantic Organicism', in *Mary Shelley: Her Circle and Her Contemporaries*, ed. by L. Adam Mekler and Lucy Morrison (Cambridge Scholars, 2010), pp. 27–43 (p. 30).

[93] Darwin, *Zoonomia*, I, p. 133.

[94] Dushane, '"Mere Matter"', para. 18.

[95] Darwin, *Zoonomia*, I, p. 132.

[96] Darwin, *Zoonomia*, I, p. 133.

[97] Darwin, *Zoonomia*, I, p. 133. Darwin does not see human identity as fixed but as an ever-changing set of ideas that people hold about themselves, observing that 'every deduction of reasoning, every sentiment or passion, with every fibre of the corporeal part

Varela puts it, 'identity amounts to a self-produced coherence'.[98] The ability to perceive time, according to Darwin, gives humans the ability to plan for the future, a capacity that plants and animals lack. This is humankind's primary difference from other creatures, who are 'almost perpetually employed about their present pleasures, or their present pains, and … seldom busy themselves about the means of procuring future bliss or avoiding future misery'.[99] In enactive terms, identity, at an individual and species level, emerges from the ways that organisms interact and shape their milieu because it is partially constituted by the 'body in motion-and-space' and partially by what it 'brings forth [as] a world from an environment'.[100]

For Darwin, emotion grounds subjectivity, and he attributes its development to interactions between innate structures of the body and those of the external world. Because it is regulated by sensorial power, emotion falls under the umbrella of cognition. Seeing it just as much a part of cognition as reason, he expands Enlightenment definitions that limit cognitive activity to ratiocination and consciousness. Even though reason is the provenance of logic and feeling states result from emotion, both originate in the body. However, as the regulator of cognitive processes, sensorial power initiates both. Tracing cognitive development back to infancy, Darwin effectively claims that emotions develop well before reason.[101]

Building on earlier work by Hartley and Priestley, which had already established a relationship between cognition and affective experiences, Darwin claims that physical sensation is the basis not only of ideas of the mind but also of emotion. Playing on the multiple connotations of *feelings*, Hartley and Priestley use the term to signify intellectual ideas *and* emotional states. According to Hartley, feeling can refer to the physical sensation of touching an object, the 'Sense of Feeling', which provides the raw material for thought, or it can denote the emotional valence that attends all ideas.[102] Hartley calls sensations and the ideas they generate 'internal Feelings of the mind', while

of our system, has been subject almost to annual mutation; while some catenations alone of our ideas and muscular actions have continued in part unchanged'. Darwin, *Zoonomia*, I, p. 133.

[98] Varela, 'Autopoiesis and Intentionality', p. 4.

[99] Darwin, *Zoonomia*, I, p. 59.

[100] Varela, 'Autopoiesis and Intentionality', pp. 13, 14.

[101] Goldstein and Noel Jackson make similar points. Goldstein contends that reason, along with volition and association, 'arises as a secondary effect, or perhaps better, an alternative description, of events first narrated under the aspect of a more basic physical configuration'. Goldstein, 'Nerve Poetry and Fiber Art'. Jackson notes that both Darwin and David Hume hold that 'sensation, passion, and imagination are active properties of reason and not exceptions to it'. Noel Jackson, 'Rhyme and Reason: Erasmus Darwin's Romanticism', *Modern Language Quarterly*, 70.2 (2009), pp. 171–194 (p. 182).

[102] Hartley, *Observations*, I, p. 117.

Priestley explains that ideas are 'terms expressing actual varieties in our mental feelings, occasioned by the impression of external objects'.[103] In their view, pain and pleasure accompany all sensation in some degree and serve as the lynchpin that connects ideas and emotions. Priestley recapitulates Hartley's view that pain and pleasure, which correlate to '*fear* and *love*', underlie every emotional state.[104] The implication is that all mental ideas are coloured by emotion, even those produced by reason. Likewise, Darwin associates perception with pleasure and pain. *Temple* describes the abstract process by which an organism's interactions with its environment engenders both feeling and thought:

> First the new actions of the excited sense,
> Urged by appulses from without, commence;
> With these exertions pain or pleasure springs.
> And forms perceptions of external things.[105]

This succinct illustration of the cognitive process represents it as recursive and interactive. Darwin starts the process with 'appulses', energetic motions that 'urge' the organism into action, as in his earlier verses that illustrate life coming into being. Although life is immanent to the 'fibre-woven frame', it requires a nudge to excite its activity.[106] Darwin minimizes the role of the external by placing the phrase 'appulses from without' on the second rather than the first line of the quatrain. This effectually directs attention towards the organism and its 'excited sense' that engender 'new actions'. However, the external is necessary to illustrate the emotional valence that from 'pain or pleasure springs' when an organism interacts with its environment, and which is a constitutive part of its 'perceptions'. This passage emphasizes not only that cognition is enactive, requiring both a percipient and a perceived, but also that perception is intimately bound up with happiness and fear, the basic emotions tied to pleasure and pain.

The ontogenetic account of emotional development that Darwin outlines in *Zoonomia* offers a more detailed account of how feeling states and their physical expressions originate in structured interactions between the embodied mind and its surroundings. Darwin challenges the dualist position by arguing that 'All emotions and passions arise out of the exertions of ... the animal sensorium'.[107] To account for emotion, Darwin traces it back to the body's

[103] Hartley, *Observations*, I, p. ii; Joseph Priestley, *Hartley's Theory of the Human Mind on the Principle of the Association of Ideas, with Essays Relating to the Subject of It* (J. Johnson, 1775), p. xl.

[104] Priestley, *Hartley's Theory*, p. xxxiii.

[105] Darwin, *Temple*, 3.2.55–58.

[106] Darwin, *Temple*, 1.4.246.

[107] Darwin, *Zoonomia*, I, p. 57.

reaction to sensory stimuli in infancy. For example, sadness and its physiological expressions begin with the shock of birth, when the 'coldness and dryness of the atmosphere, compared with the warmth and moisture, which the new-born infant had just before experienced, disagreeably affects the … lacrymal [*sic*] sack', causing tears to flow and the nose to run.[108] The abrupt transition is painful and the sensation of pain constitutes the idea that associates with the embodied state of unhappiness and grief caused by the shift from a comfortable environment to an unpleasant one. As the infant repeatedly encounters other distasteful stimuli, the idea of grief becomes habitually associated with a certain physiological response, crying. Moreover, the infant's bodily structures, in concert with its environment, ensure that this pattern will form because the 'disagreeable irritation of the lacrymal [*sic*] ducts … must frequently recur, till the tender organ becomes used to a variety of odours' and the infant adjusts to its new environs.[109] Emotions, according to Darwin, develop when structures in the body interact with compatible structures in the environment.

Emotion also develops from intersubjective experiences between the mother and child, which lay the foundations for aesthetic appreciation. Darwin's discussion implies that aesthetics is fundamental to acquiring a sense of self. Where Ruderman criticizes Darwin's account of aesthetics and subject-formation because 'there is no acknowledged sense of loss or mourning in the absence of the "original" object', I maintain that Romantic structures of engagement offer a better model than psychoanalysis for explaining Darwinian subjectivity.[110] Psychoanalysis claims that 'loss and alienation are important in that they precipitate desire and signal retrospectively the importance of the object, all of which subtend our acts of becoming a subject'.[111] However, in Darwin's account, Romantic structures of engagement in the infant interact with those in its mother or caregiver as it breastfeeds. Breastfeeding is a highly pleasurable, intersubjective experience in Darwin's representation of it that contributes to subject-formation at the same times as it lays the ground of the infant's aesthetic sensibilities. His account of subjectivity begins not in alienation and loss, but in pleasure and plenitude:

[108] Darwin, *Zoonomia*, I, p. 149.

[109] Darwin, *Zoonomia*, I, p. 150.

[110] Ruderman, *The Idea of Infancy in Nineteenth-Century British Poetry*, p. 75.

[111] Ruderman, *The Idea of Infancy in Nineteenth-Century British Poetry*, p. 75. Darwin does not address what happens to people who do not have pleasurable experiences in infancy; however, the theory of association allows for different developmental experiences that lead to different outcomes. Additionally, situating this account of subject-formation within his nosology, which includes mental disease, suggests that his theory takes psychological pathology into account. Tom Wedgwood theorizes the effect of negative infant and childhood experiences in his unpublished notebooks. See Thomas Wedgwood, 'Correspondence and Papers', V&A Wedgwood Collection, loc. E40 28478.

> Delighted by the softness and smoothness of the milky fountain ... the infant embraces with its hands, presses with its lips, and watches with its eyes; and thus acquires more accurate ideas of the form of its mother's bosom.[112]

This immersive, intersubjective experience is intensely euphoric as it engages several sensory modes, from oral to tactile to visual. It activates cross-sensory associations that link pleasure not only with the act itself but also with the world beyond the infant body. The suckling babe 'gather[s] information about objects and events simultaneously through different sensory modalities', which sharpen its undeveloped senses.[113] Perceptual acuity and growing awareness of the external world are not, in Darwin's reckoning, haunted by loss but accompanied by the alleviation of suffering (hunger) and the joy of communing with another living being. Through repetition, this euphoria matures into love, which attaches to the mother. It also associates these pleasures with the female form.

In adulthood, these feelings of pleasure recur in response to beauty. According to Darwin, in

> our mature years, when any object of vision is presented to us, which by its waving or spiral lines bears any similitude to the form of the female bosom, whether it be found in a landscape with soft gradations of rising and descending surface, or in the forms of some antique vases, or in other works of the pencil or chisel, we feel a general glow of delight, which seems to influence all our senses.[114]

Darwin's 'waving' and 'spiral lines' are reminiscent of Burke's definition of beauty as soft, smooth, and round and of William Hogarth's serpentine line of beauty.[115] This correlation between beauty and the female form also accords with Burke's assertion that the 'part of a beautiful woman where she is perhaps the most beautiful [is] about the neck and breasts; the smoothness; [and] the easy and insensible swell' of her décolletage.[116] However, Darwin goes further than Burke, suggesting that artwork engages and stimulates 'all of our senses' and that these responses originate in early relationships with other people

[112] Darwin, *Zoonomia*, I, p. 145.

[113] Peter Walker, 'Cross-Sensory Correspondences: A Theoretical Framework and Their Relevance to Music', *Psychomusicology*, 26.2 (2016), pp. 103–116 (p. 103).

[114] Darwin, *Zoonomia*, I, p. 145.

[115] See William Hogarth, *The Analysis of Beauty*, ed. by Ronald Paulson (Paul Mellon Centre for British Art by Yale University Press, 1997). I owe thanks to Richard C. Sha for pointing out this connection to me.

[116] Edmund Burke, *A Philosophical Enquiry into the Sublime and Beautiful*, ed. by Paul Guyer, Oxford World's Classics (Oxford University Press, 2015), p. 93.

and the environment. These aesthetic experiences in infancy contribute to the infant's burgeoning sense of self, as it learns to distinguish itself from its mother and its surroundings. Darwin's definition of identity as 'an idea of our own person' that changes over time implies that the subject acquires a more definite sense of self as it accrues experience.[117] Thus, the subject's growing self-awareness depends on interactions with other people. Indeed, the infant's ability to gradually recognize its physical and mental separateness from its mother does not precipitate a crisis but germinates into a subjective sense of self that becomes more defined as it matures. Rather than positing something like psychoanalytic loss, Darwin emphasizes the intense pleasure that accompanies a growing awareness of autonomy that takes its origin in intersubjectivity.

Emotional development, subject differentiation, and aesthetics also play a pivotal role in Darwin's ideas about the evolution of society. These three things make sympathy possible, which, in turn, leads to the development of language. Children develop emotion via association as well as mimetically, which teaches them to recognize it in others and to sympathize with them. Darwin claims that people 'become acquainted with the passions of others' by experiencing 'the effects of them … on our own bodies' and by assuming the bodily 'attitude that any passion naturally occasions'.[118] In both cases, observing others triggers a mimetic response that enables people to gain some sense of what others are feeling. When a parent frowns or scowls at children, they involuntarily emulate these facial expressions. As their muscles imitate these actions, they experience the feelings associated with them, and Darwin claims that this is why 'children long before they can speak, or understand the language of their parents, [are] frightened by an angry countenance'.[119] Eventually, this unconscious knowledge develops into conscious awareness of and the ability to identify psychological states based on facial expressions. According to Darwin, 'From … our aptitude to imitation, arises what is generally understood by the word sympathy'.[120] He claims that in the early history of the species, this understanding evolved into complex verbal and written language systems. Embodied emotions are 'the natural signs by which we understand each other, and on this slender basis is built all human language. For without some natural signs, no artificial ones could have been invented or understood'.[121] Basing sympathy on mimesis, Darwin foregrounds the body in his account of language and society, in contrast to earlier theories by moral philosophers. Adam Smith, for instance, claims that language 'naturally begin[s] to form'

[117] Darwin, *Zoonomia*, I, p. 134.
[118] Darwin, *Zoonomia*, I, p. 146.
[119] Darwin, *Zoonomia*, I, p. 146.
[120] Darwin, *Zoonomia*, I, p. 147.
[121] Darwin, *Zoonomia*, I, p. 147.

when people 'endeavour to make their mutual wants intelligible to each other, by uttering certain sounds, whenever they meant to denote certain objects'.[122] Where Smith emphasizes objects, Darwin stresses the body's natural signs. In *Temple*, he describes language as the capacity of 'minds to interchange the thought / And sound, the symbol of the sense', which explicitly frames it as embodied and intersubjective.[123]

Although Darwin locates humankind's symbol system within a complex of embodied and embedded sensations and perceptions that occur within a bounded set of interactions between subjects and between subjects and their milieu, he refuses to fetishize human experience. His analysis of emotion moves seamlessly between human and animal examples, frequently comparing human children to kittens and puppies, which indicates that he sees emotion as universal. Darwin takes pains to argue that language is not unique to human beings. He contends not only that animals and insects 'are possessed of some artificial language' but also that 'there is so great a resemblance in the natural language of the passions of all animals, that we generally know' their moods, 'and they have the same knowledge of us'.[124] Likewise, La Mettrie thinks that animals speak 'the language of feeling, such as moans, cries, caresses, flight, sighs, song, in a word all the expressions of pain, sadness, aversion, fear, daring, submission, anger, pleasure, joy, tenderness, etc.'[125] Diderot more sarcastically states: 'The animal is a hydraulic machine. What idiotic things can be said following this one supposition'.[126] For Darwin, intersubjectivity occurs in other species as well, with a whole range of emotions similar to those listed by La Mettrie. By making emotion rather than consciousness the basis of subjectivity, Darwin opens it up to nonhuman lifeforms. Unlike some *Naturphilosphen*, he has no investment in establishing a transcendental subject. Rather, he distributes subjectivity across nature but not as a supersubject.

In *Zoonomia*, Darwin explicitly attributes all four stages of cognition—from irritability to associability—to both animals and plants. In the section 'Of Vegetable Animation', for example, he argues that the movement of plants cannot be reduced to a mere instinct of mechanical impulse and that they demonstrate a form of subjectivity. Darwin infers that 'the vegetable world possesses some degree of voluntary powers' because they have the 'necessity to

[122] Adam Smith, *The Theory of Moral Sentiments*, 4th edn (W. Strahan, J. and F. Rivingtong, W. Johnson, T. Longman, and T. Cadell, 1774), p. 437, Eighteenth Century Collections Online <link.gale.com/apps/doc/CW0122297518/ECCO?u=asecs&sid=bookmark-ECCO&xid=f2003db8&pg=1> [accessed 2 May 2008].

[123] Darwin, *Temple*, 3.7.396–397.

[124] Darwin, *Zoonomia*, I, pp. 153, 158.

[125] Mettrie, *Machine Man*, p. 150.

[126] Diderot, *Interpreter of Nature*, p. 135.

sleep', which he defines as the 'temporary abolition of' the power of choice.[127] He sees actions such as curling their tendrils or moving to face the sun not as the function of irritability but as an act of will. With the same understanding of freewill as Hartley, Darwin holds that even nonhuman lifeforms, including plants, have a degree of choice within the constraints of natural law. Similarly, he claims that when plants are injured, their response is 'not owing to any mechanical vibrations' but rather to an awareness that they have been hurt.[128] In the third edition of *Zoonomia*, Darwin makes this point more unequivocally, adding that they react only after the 'plant seem[s] sensible of the injury'.[129] In response to the question of 'whether vegetables have ideas of external things', Darwin answers yes.[130] Tracing the ingress of sensation by comparing plants to animals, he concludes that not only do they have a 'common sensorium' but that they 'consequently possess ideas of so many of the properties of the external world, and of their own existence'.[131] With many of the same characteristics of animals and human beings, plants, in Darwin's estimation, possess subjectivity and a degree of consciousness.

While in *Zoonomia* Darwin insists that plant reproduction is not mechanistic but driven by 'the passion of love', in *Loves of the Plants*, his first long poem, he offers an extensive representation of the inner life of life that illustrates subjectivity by dramatizing the science in a way that his readers can relate to.[132] The poem presents the Linnaean system of plant reproduction as a boudoir drama that personifies plants as 'Beaux and Beauties [that] crowd the gaudy groves, / And woo and win their vegetable Loves' in order to get readers to look at the natural world differently.[133] Darwin anthropomorphizes plants not only, as Dahlia Porter suggests, in the hope that the 'allegory of pistils and maids might, then, speak the truth of human nature' but also to demonstrate his conviction that plants experience a range of affects and motives that are consistent with their physical organization.[134] Where Porter contends that the literalization of metaphor reveals the danger of Darwin's 'rhetorical figures', which 'threaten to destabilize the umbrella analogy between the laws

[127] Darwin, *Zoonomia*, I, p. 103.

[128] Darwin, *Zoonomia*, I, p. 104.

[129] Erasmus Darwin, *Zoonomia; Or, The Laws of Organic Life*, 3rd edn, 4 vols (J. Johnson, 1801), I, p. 141.

[130] Darwin, *Zoonomia*, I, p. 105.

[131] Darwin, *Zoonomia*, I, p. 107.

[132] Darwin, *Zoonomia*, I, p. 105.

[133] Erasmus Darwin, *The Loves of the Plants*, in *The Botanic Garden: A Poem in Two Parts. Part II. Containing The Loves of the Plants. A Poem. With Philosophical Notes*, 3rd edn (J. Johnson, 1795), 1.9–10.

[134] Dahlia Porter, *Science, Form, and the Problem of Induction in British Romanticism*, Cambridge Studies in Romanticism, 120 (Cambridge University Press, 2018), p. 93.

of nature and the operation of the mind by revealing it as a trope', I argue that the figures in *Loves* are meant rather more literally than Darwin's humorous verse initially indicates.[135] His personification of plants is not merely a poetic device but a way to make more visible his levelling theory of cognition.[136] As Lawrence Buell argues, his goal is to 'put nonhuman species and communities on the same footing as the human'.[137] Darwin figuratively reveals the inner lives of plants by likening their reproduction to human rituals of courtship, love, and sex. While recounting the jealousies, triumphs, and tragedies of the vegetable kingdom, Darwin attributes desire, love and grief, 'horror', 'rage and hate', and delight to plants as they fulfil their immanent purpose to thrive and procreate.[138]

By imaginatively representing the inner lives of plants, Darwin not only tries to generate sympathy for nonhuman lifeforms, but he also lays the groundwork for his organicism, which he develops more thoroughly in *Zoonomia* and *Temple*. In *Loves*, his plants recruit readers' sympathies (or outrage) by acting out stories from Greek and Roman mythology, medieval romance, and other familiar topoi. His tableaux cast the nonhuman world as gripped by intense emotions that that are legible to human beings. In one example, he represents freshwater algae as a passionately loving couple separated by a storm. The conflict hinges on whether they will be reunited or estranged forever. Darwin endows these hairy clumps of seaweed with 'glistening eyes' and 'firm arms', a tactic that helps readers envision them as feeling beings.[139] He also gives Ægea a voice to express her fears and relief. Upon seeing her swain swimming towards her, she falls to the ground, crying '—"'Tis he, 'tis he!---my Lord, my life, my love!' then 'meets her refluent lover in the waves'.[140] This brief but passionate love story implies that plants are capable of fidelity, strength, and bravery. Other plants exhibit equal, if not more sensational, psychological spectacles, including the Oedipal conflict that Clematis Nonus acts out during its reproductive cycle.[141] Although Darwin's narrative is fanciful, he connects nature and humans by common feeling.

In addition to personifying plants, Darwin further blurs the boundary

[135] Dahlia Porter, 'Scientific Analogy and Literary Taxonomy in Darwin's *Loves of the Plants*', *European Romantic Review*, 18.2 (2007), pp. 213–221 (p. 218), doi:10.1080/10509580701297950.

[136] For a discussion of Darwin's use of personification, see Packham, *Eighteenth-Century Vitalism*, pp. 152–156 and Porter, *Science, Form, and the Problem of Induction*, pp. 98–102.

[137] Lawrence Buell, *The Environmental Imagination: Thoreau, Nature Writing, and the Formation of American Culture* (Belknap Press of Harvard University Press, 1995), p. 186.

[138] Darwin, *Loves*, 4.78, 1.70, 2.287, 3.133, 4.148.

[139] Darwin, *Loves*, 4.394, 4.391.

[140] Darwin, *Loves*, 4.395, 4.400.

[141] Darwin, *Loves*, 1.125–138.

between human beings and other lifeforms based on their capacity for ratiocination. He challenges the Cartesian assumption that only humans can reason by claiming that their achievements are not unique. Defining reason as the ability to entertain 'an idea of the dissimilitude of any two ideas' by voluntarily calling them to mind and comparing them, Darwin contends that insect colonies, for example, show evidence of it.[142] He asserts that 'If we were better acquainted with the histories of those insects that are formed into societies, as the bees, wasps, and ants, ... we should find, that their arts and improvements ... arose in the same manner from experience and tradition, as the arts of our own species'.[143] He offers observational evidence, 'which fell under [his] own eye, and shewed the power of reason in a wasp, as it is exercised among men'.[144]

Darwin further asserts 'that animals can enter into contracts' as evidenced by 'contracts of friendship with each other, and with mankind' and the 'tacit contract' between young animals that 'they will not hurt each other' when they play.[145] The relationships between people and their dogs are almost Lockean, in his view, because they resemble 'all contracts [that] are made among men, that do not understand each others [*sic*] arbitrary language'.[146] As with emotion and other cognitive activity, physiological structure shapes capacity. Thus, with less complex central nervous systems, animals reason about fewer ideas than humans do. Despite this difference in degree, however, human beings have more characteristics in common with other lifeforms than differences that make them unique. Darwin's refusal to sharply separate human drives 'from animal pleasures, "vegetative" neurological responses, and sympathetic and parasympathetic processes' does not degrade subjectivity, as Ruderman contends.[147] Rather, Darwin undermines the Cartesian separation of humankind from the natural world by eliminating the insurmountable division based on mind and spirit. His natural theory of the social contests the notion that society is a divine gift that sets humans above the rest of the natural world.

Posthuman Organicism

Across his body of work, Darwin emphasizes the similitude between nonhuman creatures and human beings, rather than their differences. In fact, he does more than merely blur the boundaries. Where Bewell sees Darwin's ostensible 'tribute to nature' as 'a paean to human industry and invention', I argue that

142 Darwin, *Zoonomia*, I, p. 46.
143 Darwin, *Zoonomia*, I, p. 178.
144 Darwin, *Zoonomia*, I, p. 183.
145 Darwin, *Zoonomia*, I, p. 170.
146 Darwin, *Zoonomia*, I, p. 79.
147 Ruderman, *The Idea of Infancy in Nineteenth-Century British Poetry*, p. 77.

Darwin takes humans down a notch with an account of life and mind that challenges Cartesian speciesism.[148] His levelling of the great chain of being integrates humankind into the natural order and posits radical relationality among the living. What is more, Darwin advances a model of organicism that puts animate and inanimate nature into mutual co-dependence. This commitment to interdependence informs his social vision, which according to Priestman was consistent with other revolutionary democrats of the day.[149] Like his friend Thomas Beddoes, he called for advancements in medicine to improve the quality and longevity of life in all classes of people and advocated more sympathetic attitudes towards and treatment of other people and other lifeforms. However, these positions were at odds with his role as a leader in the Industrial Revolution and Midlands Enlightenment. I agree with Bewell that even as he represents nature as omnipotent, Darwin advances a 'thoroughly modern concept of nature ... that was undergoing ceaseless change and transformation [that was] inescapably bound up with global commerce, industry, and consumption'.[150] Darwin's work remains silent about the nascent effects of industrialization, such as the effluvium generated by new technologies—the coal-burning steam engine, for example.[151] Whether his ignorance stems from naiveté or wilful negligence, Darwin's scientific progressivism undermines his vision of nature as alpha and omega. Even so, I maintain that his posthuman organicism offers something important to twenty-first-century readers because it contests Enlightenment humanism and stands as a warning. Darwin advocates an ethics of relationality predicated on humans' recognition that we are embedded in an environment that shapes us just as much as we shape it, which provides a counterview to the Cartesian metaphor that abides even today.

Darwin's evolutionary theory displaces human beings from the centre of the narrative and brings their debt to other species to the fore. Darwin claims that 'the whole is one family of one parent', indicating that humans owe their sophisticated cognitive abilities to earlier lifeforms. He also attributes the origin of society to simpler organisms, claiming that it developed long before humankind.[152] Darwin overturns Christian anthropological accounts of history by claiming that the world is much older than the Bible indicates and by asserting that human beings were not around for much of that history. In *Zoonomia* he asks, 'would it be too bold to imagine, that in the great length of time, since the earth began to exist, perhaps millions of ages before the

[148] Bewell, 'Erasmus Darwin's Cosmopolitan Nature', p. 31.

[149] Priestman, *The Poetry of Erasmus Darwin*, p. 193.

[150] Bewell, 'Erasmus Darwin's Cosmopolitan Nature', p. 40.

[151] Thanks to Benjamin Colbert for this point.

[152] Darwin, *Zoonomia*, I, p. 1.

commencement of the history of mankind ... that all warm-blooded animals have arisen from one living filament?'[153] In a discussion with Holbach, reported in a letter to a friend, Diderot made similar claims about human evolution. As 'natural productions of the soil', humankind is subject to the same slow developmental process as all other creatures, and existed in the early epochs of the Earth, 'but not as he is. First, a certain something; then another certain something; and then, after several hundreds of millions of years and so many more certain somethings, the bipedal animal who has the name of man' appears.[154] While not as detailed as Darwin's account, it is clear that Diderot too sees human beings as originating in simpler lifeforms. While Darwin touches on many of these ideas in *Zoonomia*, he versifies them in *Temple*, which narrativizes the creation of the Earth and the *Origin of Society*, the poem's subtitle. His last poem, published posthumously, *Temple* depicts Darwin's posthuman organicism more clearly than his other published works.

Before he wrote *Temple*, Darwin started a poem that he never finished entitled *The Progress of Society* that foregrounds human inventions. The surviving fragments of this unfinished work reveal that instead of organizing 'progress' according to element, as in *Economy*, he intended to categorize them by stadial theory's four phases of social evolution—hunting, pastorage, agriculture, and commerce, and to end with a utopian golden age.[155] According to the stadial view of history, the development of culture is seen as a 'reciprocal relationship ... between history and mental development'.[156] In this view, phylogeny follows ontogeny. Just as a human infant acquires more sophisticated language and motor skills as it matures, culture also progresses over time.[157] It seems that *Progress* would have represented the 'modern, cosmopolitan nature, whose character would be determined by science and the marketplace' that Bewell claims Darwin advances in *Economy*, but Darwin abandoned *Progress* to write *Temple*, which situates inventions and human society within a much larger and longer narrative about the origin of the cosmos.[158]

Porter observes that in contrast to the 'personified maids and swains [in *Loves*], personifications in *Economy* are abstract concepts, often acting as invisible agents behind the scenes'.[159] Similarly, in *Temple* the nonhuman agents

153 Darwin, *Zoonomia*, I, p. 505.

154 Qtd in Crocker, *Diderot's Chaotic Order*, p. 25.

155 See Priestman, *The Poetry of Erasmus Darwin*, Appendix A, pp. 260–282 for a reconstruction of *The Progress of Society* from manuscript notebooks.

156 Ann Wierda Rowland, *Romanticism and Childhood: The Infantilization of British Literary Culture*, Cambridge Studies in Romanticism, 93 (Cambridge University Press, 2015), p. 43.

157 See Rowland, *Romanticism and Childhood*, pp. 41–43, for an account of how stadial theory and psychology intersect in Enlightenment conjectural history.

158 Bewell, 'Erasmus Darwin's Cosmopolitan Nature', p. 40.

159 Porter, *Science, Form, and the Problem of Induction*, p. 103.

act long before humans come into being. Humankind does not occupy centre stage in this (r)evolutionary poem that locates society's origins not in Eve and Adam but in single living filaments.[160] The narrative commences with the dawn of time, thus temporally connecting the origin of society with the origin of life by situating the former within the story of evolution. After Darwin establishes how the Earth was formed, his depiction of life emerging from matter represents the inanimate and animate parts of nature in a mutually constitutive relationship. Driven by immanent purposiveness, the first lifeforms strive to 'maintain and enhance their patterns of life' by seeking out food, as discussed above.[161] Darwin illustrates the reciprocity of this process, as organisms literally incorporate nonorganic elements that enable them to create and maintain the constituent parts of their living bodies. All creatures, he claims, have 'powers to arrest the winds and floods' so that they might 'To bones convert them, or to shells, or woods'.[162] In this condensed image, living filaments ingest particles in their environment and 'convert them' into skeletal systems (whether an endoskeleton, exoskeleton, or the rigid cell walls of a plant) vis-à-vis metabolic processes. In Darwin's vision, evolution 'occurs on both sides (of the organism and the environment). Organism and environment mutually determine each other through a set of historical networks, so that they cannot exist without each other'.[163] Where Cartesians imagine humankind divorced from nature, Darwin forecloses that possibility by showing life evolving in tandem with its physical environment. Both the animate and inanimate aspects of nature adapt to meet constantly changing structural conditions that result from their interactions.

According to Darwin, the evolution of the environment depends as much on the organisms it supports as they depend on it. As they live and die, animate beings excrete waste materials that the landscape incorporates into itself. Before the emergence of life, the Earth was relatively formless in Darwin's account. It consisted of 'sun-beams' and 'primeval caves' on a 'shoreless earth'.[164] The image portrays a sky parted from the sea, but without land or plant life.

[160] While some scholars consider Erasmus Darwin's theory proto-evolutionary, I see his single living filaments as comparable to single-celled organisms. Even though his science was wrong, his paradigm is effectively evolutionary. Despite defensive claims 'that his grandfather's thinking had not had "any effect" on his own', Charles Darwin was more influenced by his grandfather's work than he cared to admit. Devin Griffiths, 'The Fertile Darwins: Epigenesis, Organicism, and the Problem of Inheritance', *Romanticism on the Net*, no. 66–67 (2016), para. 1, doi:10.7202/1069959ar.

[161] Ellis, *Curious Emotions*, p. 180.

[162] Darwin, *Temple*, 1.4.265–266.

[163] Jacek Seweryn Podgórski, 'From Autopoiesis to Natural Drift Metaphor', *Ecological Questions*, vol. 13 (2014), pp. 81–87, p. 85.

[164] Darwin, *Temple*, 1.4.233, 1.4.232.

Directly contradicting the *Genesis* story, wherein God created the sea and land before he populated it with creatures, in Darwin's account, 'Organic Life began beneath the waves', substantially altering the environment as it evolved.[165] Secreting waste while alive and decomposing after death, these early sea creatures, Darwin claims, formed 'the vast beds of argil, lime, and sand / And from diminish'd oceans form the land!'[166] In this scene, land emerges in consequence of the ocean inhabitants' life processes. This in turn creates an environment hospitable to amphibious and land-dwelling creatures. In the footnote that accompanies these lines, Darwin confirms 'that the exertions of vegetable and animal life convert the fluid parts of the globe into solid ones'.[167] In *Zoonomia*, he elaborates, explaining that the nonorganic elements 'of the earth ha[ve] been formed out of organic recrements', that is, 'the secretions of organic life', and the 'immense beds of limestone, chalk, marble' are formed 'from the shells of fish; and the extensive provinces of clay, sandstone, ironstone, coals' come 'from decomposed vegetables'.[168]

Darwin was not the only one who saw animal remains in this light. Coral reefs 'formed an important link between inert and living worlds', and the existence of 'coral islands, discovered on the Cook voyages, was thought to explain the origins of the lands of the world'.[169] John Walker, a Professor of Natural History at the University of Edinburgh, remarked in a lecture that 'coral shoots from the bottom from a very considerable depth and grows gradually upward till it reaches the surface of the sea ... gathers the exuviae of the ocean, and the seeds of plants ... acquires soil, and in this way there can be a regular progression traced between a single plant of coral and an extensive island filled with plants and animals'.[170] Walker 'does not give any indication of how long this process takes', using 'human artefacts and written histories to date strata' because he believed the Earth was only a few thousand years old.[171] In contrast, Darwin saw the formation of land as necessary to support the evolution of life, a much longer process. As the landmass increases in area,

[165] Darwin, *Temple*, 1.4.234.

[166] Darwin, *Temple*, 1.4.267–268.

[167] Darwin, *Temple*, n. 1.4.268.

[168] Darwin, *Zoonomia*, I, p. 509.

[169] Graham Durant and W. Ian Rolfe, 'William Hunter (1718–1783) As Natural Historian: His "Geological" Interests', ed. by Hugh Torrens, *Earth Sciences History*, 3.1 (2007), pp. 9–24 (p. 15), doi:10.17704/eshi.3.1.6451t601508g3541.

[170] John Walker, *Lectures on Geology: Including Hydrography, Mineralogy, and Meteorology, with an Introduction to Biology*, ed. by Harold W. Scott (University of Chicago Press, 1966), p. 179.

[171] Matthew Daniel Eddy, 'Geology, Minerology and Time in John Walker's University of Edinburgh Natural History Lectures (1779–1803)', *History of Science*, 39 (2001), pp. 95–119 (p. 108).

it provides a suitable environment for life to further complexify and diversify. As living organisms become more structurally sophisticated, so does the Earth.

The relationship that Darwin envisions between living creatures and their environment fits within a progressive narrative. The interdependence that he posits does not merely maintain the system but improves it. When organic bodies die, their decay replenishes the nonorganic world by releasing gases into the atmosphere and by returning chemicals and matter to the soil. They replace what they use up during their lifetimes, but they also enrich the environment, which in turn creates hospitable conditions that sustain and improve subsequent generations of organisms. Darwin notes that because the 'remains of former life are not again totally decomposed, or converted into their original elements, they supply more copious food to the succession of new animal or vegetable beings' by providing 'materials convertible into nutriment with less labour or activity of the digestive powers'.[172] In the cycle of life and death, previous generations make survival easier for the species that come after them. With improved living conditions and more easily digestible food, 'the quantity or number of organized bodies, and their improvement in size ... has been continually increasing' over the course of natural history.[173]

Advancing a concept of organic unity in which the structuring principle does not and cannot exist separately from its instantiation, Darwin offers a robust model of Romantic organicism freed from transcendentalist trappings. In *Temple*, Darwin metaphorizes the relationship between animate and inanimate nature as a tightly constructed natural system to illustrate the relationships that structure his organicism. His image of life as warp and woof suggests that it is part of a larger assemblage composed of several types of forces that work together to create and sustain the physical world. Much as fabric is made of interwoven threads, life obtains as it interweaves with the raw materials of inanimate nature. This image of life threading through 'Nature's loom' not only invokes the 'fibre-woven body' that, according to Darwin, makes life possible, it also suggests that, contrary to dualist claims, life requires embodiment.[174] The structural interdependence between the animate and inanimate reveals that relationality, reciprocity, and mutual co-dependence are the governing principles of Darwin's model of organicism.

Darwin's model also proposes a solution to the idealist problem of reconciling the absolute—the unconditioned, imperishable whole—with the finite and conditional parts that comprise it by incorporating death into his system.[175]

[172] Darwin, *Temple*, n. 4.3.453.

[173] Darwin, *Temple*, n. 4.3.453.

[174] Darwin, *Temple*, 1.4.252; Hisaoy Ishizuka, 'Enlightening the Fibre-Woven Body: William Blake and Eighteenth-Century Fibre Medicine', *Literature and Medicine*, 25.1 (2006), pp. 72–92 (p. 76), doi:10.1353/lm.2006.0020.

[175] Keren Gorodeisky, '19th Century Romantic Aesthetics', in *The Stanford Encyclopedia of*

According to Greg Ellermann, idealists could not ground the absolute in the concept of life because, when it is instantiated, its 'entwinement with death and disease threatens too strongly the natural order that any absolute is supposed to ground'.[176] However, Darwin's conception of death substitutes the immortality of matter for the eternal absolute. Diderot held a similar view, believing that matter and motion exist in an eternal cycle of structuration and dissolution.[177] Darwin had read Antoine Lavoisier's *Elements of Chemistry* (1789), which details the law of conservation of mass. In *Temple*, he uses this law to characterize death as the dissolution of matter into an undifferentiated mass (most likely understood as potential energy):

> When Time's cold hands the languid senses seize
> Chill the dull nerves, the lingering currents freeze;
> Organic matter, unreclaim'd by Life,
> Reverts to elements by chemic strife.[178]

Understood in terms of autopoiesis, once an organism can no longer execute the internal processes that enable it to maintain its organization—whether through accident or old age—its boundaries dissolve and it becomes indistinguishable from the elements that comprised it.

For Darwin, because life is not an immaterial principle superadded to the body it lives on, not in a future state but in the organism's offspring. Thus, in 'Reproduction' he asserts, 'New Life rekindles, ere the first expires'.[179] Rather than envisioning the lifeforce as unique to each animate body, Darwin generalizes it, much as gravity is a global force, as the common property of living structures. Thus, by producing offspring before it dies, each generation of species 'With finer links the vital chain extends, / And the long line of Being never ends'.[180] The perishability of matter, which poses insurmountable problems to transcendentalist organicism, becomes part of the eternal cycle that continuously renews and sustains itself, thanks to the reciprocity between the living and non-living. Darwin's organicism thus balances temporality with timelessness by providing a way to think abstractly about processes without divorcing them from materiality. Relationality abides even as its manifestations

Philosophy, ed. by Edward N. Zalta and Uri Nodelman, Fall 2016 (Metaphysics Research Lab, Stanford University) <https://plato.stanford.edu/archives/fall2016/entries/aesthetics-19th-romantic/> [accessed 1 June 2021].

176 Greg Ellermann, 'Late Coleridge and the Life of Idealism', *Studies in Romanticism*, 54.1 (2015), pp. 33–56 (p. 39).

177 Crocker, *Diderot's Chaotic Order*, p. 26.

178 Darwin, *Temple*, 2.1.5–8.

179 Darwin, *Temple*, 2.1.13–14.

180 Darwin, *Temple*, 2.1.19–20.

are constantly in flux. Exclaiming 'Live but to die, and die to revive!', Darwin sees cycles rather than forms as eternal.[181]

The principle of relationality also underwrites Darwin's ideas about the evolution and progress of society. In *Temple*, presenting the origin of society as the story of evolution, Darwin does not introduce humankind into the narrative until about a hundred lines into the third canto. The absence of humans from the first half of the poem reinforces the value of other lifeforms.[182] 'Reproduction', the process by which an organism 'Forms from fine glands another like itself', establishes primitive social formations, serving as the first 'vital chain [that] extends' into the second and third chains of society.[183] Given that the first lifeforms, according to Darwin, were nothing more than lines and tubes, society is not a human invention. As with human social formations, he claims that the first societies emerged from emotion. The precursors of social feeling come into being when asexual 'Reproduction of the living Ens / From sires to sons, unknown to sex commence', and

> Parturient Sires caress their infant train,
> And heaven-born Storge weaves the social chain;
> Successive births her tender cares combine,
> And soft affections live along the line.[184]

Darwin defines storge as the 'affection from the parent to the progeny [which] existed before animals were divided into sexes, and produced the beginning of sympathetic society'.[185] Despite this feeling of parent for offspring, isogamous organisms continued to live in isolation.[186] Sexual differentiation was required for organisms to leave their 'unsocial bowers'.[187] While Darwin's account of society is heterosexist, the division into two sexes seems to be less important than the development of 'Sexual Love', particularly since the first kind of sexually differentiated reproduction that Darwin details is hermaphroditic.[188] In his account, the difference between pistils and stamens is enough to 'give

[181] Darwin, *Temple*, 2.2.42.

[182] Even though Darwin alludes to historical and mythological figures in the first two cantos to illustrate points, they are not central to the narrative.

[183] Darwin, *Temple*, 2.1.27–28, 2.1.19.

[184] Darwin, *Temple*, 2.2. 62–64, 2.2.91–94. Darwin calls asexual reproduction *spontaneous vitality*. See Darwin, 'Spontaneous Vitality of Microscopic Animals', 'Additional Notes', in *Temple*, pp. 139–149.

[185] Darwin, 'Storge', in *Temple*, n. 9, p. 175.

[186] While it is accurate to call the first organisms isogamous, that is neither male nor female, Darwin imagines that the first lifeforms were male. See 'Of Generation', in *Zoonomia*, I, pp. 478–533.

[187] Darwin, *Temple*, 2.5.247.

[188] Darwin, *Temple*, 2.5.244.

Society his golden chains'.[189] Furthermore, in contrast to thinkers such as Burke, Darwin does not ground society in family structures. Where Burke locates the social passions in 'the society of sex', or heterosexual 'love' which not only 'contains a mixture of lust; [but has for] its object ... the beauty of women', Darwin's account of different modes of animal and vegetable love accommodates polyamory, as evidenced by some of the steamier scenes in *Loves*.[190]

By extending emotion to all species, Darwin defines society more broadly than most of his contemporaries, presenting a posthuman rather than liberal humanist vision of it. According to Buell, he blurs the 'traditional hierarchalizing boundaries separating the human estate from the rest of nature' and 'democratize[s] moral accountability'.[191] Darwin's poetry attempts to convince readers that humankind is neither outside nor above nature but a part of it. With stanzas that alternately laud human reason then detail the ingenuity of insects and birds, the structure of *Temple* reminds readers that humankind is part of the chain of society but not at the top of it.

Yet, Darwin's insistence that humankind must recognize that it is part of the material world does not resolve a fundamental tension between humankind's role in improving society and nature's proclivity to maintain balance, which runs throughout his work. The final canto of *Temple* opens with the suggestion that humankind needs to counteract the violence that Darwin represents as inherent in nature. After bemoaning 'How few, alas! in Nature's wide domains / The sacred charm of Sympathy restrains!', he presents a litany of brutalities, beginning with human wars.[192] However, as with nearly all other aspects of human nature, aggression is not an exclusively human characteristic. Darwin introduces the predatory animal that 'Unknown to mercy, tears the guiltless lamb', but a few lines later he reveals that the lamb is not so 'guiltless' after all, for it wreaks havoc in the plant world when it 'Crop[s] the young herb, or crush[es] the embryon seed'.[193] Marching through the damage wrought by insects, plants' competition for resources, and natural disasters, Darwin depicts 'the warring world' of nature as 'one great Slaughter-house'.[194] In response to this, he urges readers to view each other and 'His brother-emmets and his sister-worms' as kin.[195] This line suggests that humankind needs to mitigate violence by acting more sympathetically and charitably towards the world and its inhabitants.

189 Darwin, *Temple*, 2.2.250.
190 Burke, *Philosophical Enquiry*, p. 44.
191 Buell, *The Environmental Imagination*, p. 186.
192 Darwin, *Temple*, 4.1.2–3.
193 Darwin, *Temple*, 4.1.18, 1.4.1.22.
194 Darwin, *Temple*, 4.1.66.
195 Darwin, *Temple*, 4.3.425–28.

The second section of the canto, however, implies that good offsets evil in nature and that balance is the true character of the natural world. Following his catalogue of depredations, Darwin indexes the beautiful and good parts of nature and society. In addition to 'Zephyrs [that] cool us' and 'sun-beams [that] warm', 'fragrant flowers', 'sweet fruits', and other natural delights, Darwin lists the arts and sciences that have the capacity to 'Warm every clime, and brighten every hour'.[196] He includes social activists, such as John Howard, prison reformer; Sir Francis Burdett, a member of parliament who 'forced an investigation into Coldbath Fields prison, where arrestees were held without trial'; and Lord Moira, who fought the suspension of *habeas corpus* in Ireland.[197] Including these campaigners, who improved society through politicking, suggests that, on the one hand, active efforts to reform society are necessary to bring about greater social equity, but, on the other, that this sort human behaviour is part of the natural order.

Elsewhere in *Temple*, as well as in *Economy*, Darwin intimates that science and technology should improve nature's gifts to benefit humankind. He saw art, science, and industry as instruments of social progress with the potential to end scarcity, stop poverty, and eliminate sickness and disease. In Darwin's view, *techné* 'builds on Nature's base the works of Art' and helps create more hospitable living conditions for all creatures.[198] According to Priestman, Darwin believed technology would usher in an 'utopian future age of "philosophy" or science'.[199] However, he also recognized that technological advances have unintended consequences. For example, better medicine means reduced infant mortality and longer lifespans. Darwin identifies that improved living conditions and better nutrition would increase the population, to the detriment of the Earth:

> human progenies, if unrestrain'd
> By climate friended, and by food sustain'd
> O'er seas and soils, prolific hordes! would spread
> Erelong, and deluge their terraqueous bed.[200]

These lines show that Darwin understands that unchecked production and consumption, aided by science and technology, leads to overpopulation. Furthermore, he recognizes that these 'prolific hoards' would overrun the Earth and cause significant environmental problems. In some ways, it seems that

[196] Darwin, *Temple*, 4.2.154–156, 4.1.18.
[197] Priestman, *The Poetry of Erasmus Darwin*, p. 197.
[198] Darwin, *Temple*, 3.7.410.
[199] Priestman, *The Poetry of Erasmus Darwin*, p. 185.
[200] Darwin, *Temple*, 4.3.369–372.

Darwin foresaw the global catastrophe that defines the Anthropocene. However, following Malthusian logic, he asserts that a combination of human and natural, though mostly natural, factors—'war, and pestilence, disease, and dearth'—will restore balance by keeping the population in check.[201] Yet Darwin's belief that 'organic recrements' enrich the Earth failed to anticipate the excessive quantity and toxicity of waste that industry would generate.[202] He did not seem to understand that medicine's potential to increase life spans and technology's ability to harness nature would counteract the checks that, according to Thomas Malthus's logic, should contain the global population. Darwin also failed to predict the long-term effects of imperialism and capitalism, which would increase disparities in wealth and access to resources that he hoped to eliminate. Nor was he able to foresee the social and reproductive behaviours of people living in economically privileged societies under late-stage capitalism.[203]

Despite his commitment to science, and perhaps paradoxically, Darwin had greater faith in nature than in human technology to restore equilibrium, but his vision of natural balance is drastic. Darwin wrestles with a secular version of the problem that William King, author of *An Essay on the Origin of Evil* (1702), Gottfried Wilhelm Leibniz, and other religious thinkers, including Hartley, tried to solve, that is, how to reconcile the problem of evil if nature is benevolent and progressive. Darwin never resolves the two competing impulses that he sees at work in nature, but his model of organicism contains an extreme self-correcting mechanism. Trusting that nature's 'firm and immutable, immortal laws' will always prevail, he conjectures that nature endlessly regenerates itself.[204] Extending his theory of death to the cosmos as a whole, in *Economy* Darwin describes natural processes as 'the perpetual circulation of matter' in an endless cycle of 'growth and dissolution'.[205] Similarly, in *D'Alembert's Dream* (1830), Diderot asks, 'Who knows what races of animals have preceded us? Who knows what races of animals will come after ours? Everything changes and everything passes away, only the whole endures'.[206] Darwin takes this view further in *Temple*, wherein he claims that 'Immortal matter braves the transient storm / Mounts from the wreck, unchanging but in form'.[207]

[201] Darwin, *Temple*, 4.3.373.
[202] Darwin, *Zoonomia*, I, p. 509.
[203] Thanks to Dahlia Porter for this observation.
[204] Darwin, *Temple*, 4.3.461.
[205] Darwin, *Economy*, n. 574.
[206] Diderot, *Interpreter of Nature*, p. 72. Published under Diderot's name posthumously in 1830, *D'Alembert's Dream* initially appeared anonymously in *Correspondance Littéraire* in August 1782. Ralph H. Bowen, 'Preface', in *Rameau's Nephew and Other Works*, by Denis Diderot, trans. by Jacques Barzun and Ralph H. Bowen (Macmillan Library of Liberal Arts, 1964), pp. 89–91 (p. 90).
[207] Darwin, *Temple*, 2.1.43–44.

His notes speculate that the cycle of life-death-rebirth repeats itself not just ontogenetically and phylogenetically but also cosmically. Darwin's theory of evolution ends not at the pinnacle of human achievement but with planetary destruction. Evolution will continue, he claims,

> till the whole terraqueous sphere, and all that inhabit it shall dissolve by a general conflagration, and be again reduced to their elements.
>
> Thus all the suns, and the planets, which circle round them, may again sink into one central chaos; and may again by explosions produce a new world; which in process of time may resemble the present one, and at length again undergo the same catastrophe![208]

Temple closes with a triumphant swell that celebrates the endless regeneration of matter, proclaiming that 'Reproduction strives / With vanquish'd Death,—and Happiness survives'.[209] Driven by 'Life', or sensorial power, 'young renascent Nature conquers Time'.[210] This view suggests that if human checks fail the Earth will end and be reborn in the natural course of things. Even if it takes aeons, Darwin trusts the perpetuity of natural processes.

The value of Darwin's work in the twenty-first century is neither its moral message nor its ecological reverie but its posthuman vision of relationality among all parts of the material world. Representing 'human corporeality, in all its material fleshiness, [as] inseparable from "nature" or "environment"', Darwin's deeply anti-Cartesian view of humankind puts natural processes, rather than human beings, at the centre of the narrative.[211] He asserts neither 'a "masculinist" naturalist code' nor 'a "sentimentalist" naturist code'.[212] Rather, he represents the relationship between humans and their world as a posthuman 'space in which the human actors are still there but now are inextricably entangled with the nonhuman'.[213] Darwin invites us to 'combat the persistent nature / culture, body / mind dualisms of Western culture' by calling humankind to understand itself as embodied, more animal-like, and to understand nature as more alive than in the Cartesian worldview.[214]

Darwin offers a perspective that challenges Enlightenment speciesism and Cartesian separation and opens an ethical space to consider how we orient ourselves and act towards the nonhuman world. He reminds us not to become

[208] Darwin, *Temple*, n. 4.3.453.
[209] Darwin, *Temple*, 4.3.451–452.
[210] Darwin, *Temple*, 4.3.453–454.
[211] Alaimo, 'Trans-Corporeal Feminisms', p. 238.
[212] Buell, *The Environmental Imagination*, p. 199.
[213] Andrew Pickering, *The Mangle of Practice: Time, Agency, and Science* (University of Chicago Press, 2010), p. 26.
[214] Alaimo, 'Trans-Corporeal Feminisms', p. 257.

complacent about the need for ethical action, insisting 'That man should ever be the friend of man, / Should eye with tenderness all living forms' in order to create a more equitable society.[215] His organicism offers an alternative metaphor that can shift our understanding of ourselves in relationship to the world, but his corpus also warns us not to ignore how 'the ways in which we have interacted with the physical world' have remade 'human nature and [created] an altered natural order'.[216] His short-sightedness about capitalism, industry, and empire challenges us to really see the world we have mutually co-produced through interacting with our surroundings while under the sway of Cartesian ideology. Those of us living in the Anthropocene, under late-stage capitalism, and with the threat of fascism must understand how we have collectively created our *umwelt*.

Eventually, Darwin bowed to economic and social pressure and painted over the motto *e conchis omnia* on his carriage doors, but that does not invalidate his challenge to the view of nature as humankind's 'Other'. In his works, he metaphorically reunites a universe broken into discrete parts and dares his readers, even today, to re-evaluate their humanistic worldviews and to cultivate an ethics of relationality that connects them to the rest of the living and non-living world and, hopefully, inspires more principled action.

[215] Darwin, *Temple*, 4.3.425–426.

[216] Raymond Williams, *Problems in Materialism and Culture: Selected Essays* (Verso, 1980), p. 84.

Chapter Three

Phenomenological Organicism

Samuel Taylor Coleridge

Samuel Taylor Coleridge's first conversation poem, 'The Eolian Harp' (1796, 1797, 1817), begins as a love poem that quickly wanders into speculation about the relationships that exist among the mind, the body, the world, and God. Over the twenty-two-year revision period, the poem, according to some critics, ended up 'riddled with [thematic] inconsistencies, which ... are intellectually intolerable'.[1] Yet the poem opens with the comfortable materiality of Coleridge's relationship with Sara Fricker, and it ends there as well—no small fact, but one that has been overlooked by most critics. The opening and closing frames that bound the speaker's speculations exemplify the role of the body in Coleridge's theory of mind and imagination—inconspicuous yet crucial. The poem itself embodies his model of organicism, predicated on his opposition to Cartesianism, as it contemplates 'all of animated nature' and how its 'diversely framed' parts relate to each other.[2] Briefly positing that everything is connected via vital cognitive processes—'one vast intellectual breeze'—the poem replicates the ontological and epistemological conundrums with which he wrestled over the course of his career.[3] Furthermore, the ostensible inconsistencies in 'The Eolian Harp' enact Coleridge's aesthetic methodology, in which poetry disciplines the mind and body into habits of active engagement that enable the reader to discover divine truth by interacting with the physical world.

Eighteenth- and nineteenth-century positions on cognition and the lifeforce

[1] Ronald C. Wendling, 'Coleridge and the Consistency of "The Eolian Harp"', *Studies in Romanticism*, 8.1 (1968), pp. 26–42 (p. 26).

[2] Samuel Taylor Coleridge, 'The Eolian Harp: Composed at Clevedon, Somersetshire', in *Poetical Works*, ed. by J.C.C. Mays, The Collected Works of Samuel Taylor Coleridge, Bollingen Series, 2 vols (Princeton University Press, 2001), i, pp. 231–235 (ll. 44–45).

[3] Coleridge, 'Eolian Harp', l. 47.

were largely divided along ontological lines. In the eighteenth-century materialist camp, men of science such as David Hartley, Joseph Priestley, Erasmus Darwin, John Thelwall, and Tom Wedgwood theorized that mental activity is an organic process inherent in organized forms of matter capable of supporting life. Each of these men had a significant impact on Coleridge's thinking in the early part of his career. Hartley's *Observations on Man* (1749) provided him with a model of synthetic philosophy that incorporated God into its system of mind–body–world, which helped set Coleridge on the path towards formulating an integrated philosophy that included religion. As a Unitarian minister and man of science who saw the French Revolution in millenarian terms, Priestley appealed to a young, enthusiastic Coleridge, especially with respect to pantisocracy.[4] Coleridge's long poem *Religious Musings* (1796) offers a paean to Hartley and Priestley, along with Isaac Newton and Benjamin Franklin, all of whom he believed were ushering in a new golden age with their scientific discoveries. Although unsettled by Darwin's atheism, Coleridge admired his encyclopaedic knowledge, and even though he claimed to despise his verse some critics think it inspired his interest in philosophical poetry; he also corresponded with Thelwall and Wedgwood about Darwin's neurophysiological speculations. These men considered life as an emergent property of certain material configurations, an idea that Coleridge adopted in the 1790s and early 1800s and which influenced him long after he abandoned materialism.

In the dualist camp, Common Sense philosophers Dugald Stewart, Thomas Reid, and Thomas Brown and medical men such as John Abernethy held the position that cognition and life are attributable to an immaterial force superadded to the body that exists independently of instantiation, while idealists such as Bishop Berkeley or the many *Naturphilosophen* believed that the world is the product of supernormal cognizing activity, whether a transcendent force or God. Just as Coleridge sorted through these myriad attempts to resolve the mind–matter question, many critics have also tried to isolate his position from those of his influences. Typically, some of these influences have been preferred

[4] Coleridge and Robert Southey planned to move to Pennsylvania to set up a utopian community, based on democratic and unitarian ideals, on the banks of the Susquehanna River. Priestley, who had emigrated to America to escape persecution in Britian, inspired their plan, at least in part. Many critics consider the influence of Hartley and Priestley on Coleridge. See, for example, Peter Cheyne, '"Wild Activity of Thoughts": Associationism and the Idea of Life', *Coleridge Bulletin: The Journal of the Friends of Coleridge*, n.s., 52 (2018), pp. 17–26; Ashley Miller, '"Striking Passages": Memory and the Romantic Imprint', *Studies in Romanticism*, 50 (2011), pp. 29–53; Shelley Trower, 'Nerves, Vibration and the Aeolian Harp', *Romanticism and Victorianism on the Net*, 54 (2009); Michael Raiger, 'The Intellectual Breeze, the Corporeality of Thought, and the Eolian Harp', *Coleridge Bulletin: The Journal of the Friends of Coleridge*, 20 (2002), pp. 76–84; and Jerome Christensen, *Coleridge's Blessed Machine of Language* (Cornell University Press, 1981).

to others. For example, early-to-mid-twentieth-century critics such as Thomas McFarland, Kathleen Coburn, and Basil Willey, among others, had such a distaste for British materialism that they dismissed his engagement with these sources as a mere dalliance. Later in the twentieth century, New Historicist critics aligned Coleridge (and his brother poets) with German transcendentalism or, in Jerome McGann's phrase, the Romantic ideology.

Even though Coleridge's rejection of Cartesianism is well known, in trying to bring him in line with more reputable influences (that is, German and Greek models) McFarland and Coburn inadvertently transform him into a dualist. More recently Charles I. Armstrong, Gavin Budge, and Suzanne Webster, although they appropriately hedge their claims, brand him a dualist explicitly. According to Armstrong, despite the attempt to advance a monist ontology in 'Theory of Life' Coleridge's organicism ultimately 'yields to dualism', while Budge tries to avoid ontology by deeming him an epistemological dualist.[5] Arguing that digestion serves as an analogy for Coleridge's account of cognition, he contends that 'indigestion indicated a fundamentally dualistic relationship between mind and body', a claim that aligns him with René Descartes, albeit unintentionally.[6] At the same time, Webster claims that once Coleridge broke the 'holistic bond' with nature, his 'formulation of this dualistic relationship [between nature and God] could range from moderate to strongly radical'.[7] Yet Coleridge's opposition to Cartesian dualism make these assessments of his ontological commitments unsatisfactory.

David Vallins observes that Coleridge 'wishes to be at once a monist and dualist', but because of the fundamental incompatibility of these two positions he fails.[8] However, despite a degree of nuance, the critical consensus seems to be that Coleridge's attempts to overcome Cartesianism, like his 'poetic theory', exhibit a 'mentalistic and transcendental bias', an idea that extends to his theory of organicism.[9] Seamus Perry notes that the propensity to privilege 'the visionary' over the real 'continues to dominate most conceptions of Coleridge'.[10]

[5] Charles I. Armstrong, *Romantic Organicism: From Idealist Origins to Ambivalent Afterlife* (Palgrave Macmillan, 2003), p. 63.

[6] Gavin Budge, *Romanticism, Medicine and the Natural Supernatural: Transcendent Vision and Bodily Spectres, 1789–1852* (Palgrave Macmillan, 2012), p. 85.

[7] Suzanne Webster, *Body and Soul in Coleridge's Notebooks, 1827–1834: 'What Is Life?'* (Palgrave Macmillan, 2010), pp. 12–13.

[8] David Vallins, *Coleridge and the Psychology of Romanticism: Feeling and Thought* (Macmillan, 2000), p. 136.

[9] Alan Richardson, *British Romanticism and the Science of the Mind*, Cambridge Studies in Romanticism, 47 (Cambridge University Press, 2001), p. 65.

[10] Seamus Perry, *Coleridge and the Uses of Division*, Oxford English Monographs

Likewise, David Fairer contends that Romantic organicism more generally has long been misunderstood as standing for a teleological 'metaphysical Idea' that imposes a 'unified structure', whether onto a poem or the world.[11] Armstrong stands as a recent example of such a view, except rather than reinforcing the Romantic ideology he shows how it fails. He argues that Coleridge's struggle to reconcile the principles of unity and multeity meets the same end as Immanuel Kant's model—that is, by prioritizing the unifying impulse it ultimately values hierarchy over interdependence.[12] Armstrong, like some earlier critics, analyses Coleridge's theory without attending to its biological influences or considering its relationship to embodiment.[13] Perry and Fairer, on the other hand, challenge scholarship that attends to the transcendental at the expense of the physical. Perry acknowledges Coleridge's irresistible attraction to the material world, citing it as a cause of his divided philosophy, while Fairer grounds Coleridge's definition of the organic in the relationships he had with the other poets in his circle. Along similar lines, drawing on historical scientific and philosophical debates in which he participated, I locate Coleridge's concept of organicism and his aesthetics in the physical rather than the metaphysical.

My approach considers the scholarship that traces Coleridge's ideas to his extensive knowledge of Romantic medicine and his fraught experience of embodiment. As Neil Vickers notes, 'Coleridge was an attentive student of ... speculative, neuropathological medical theories'.[14] Kevis Goodman contends that Coleridge's investments in form and metre owe much to notions of 'health', which 'was thought to depend on an equilibrium between forces coming from the external and motions working internally within the animal economy'.[15] Where she traces the influence of William Cullen's and Robert

(Clarendon Press, 1999), p. 38.

[11] David Fairer, *Organising Poetry: The Coleridge Circle, 1790–1798* (Oxford University Press, 2009), p. 10.

[12] Armstrong, *Romantic Organicism*, p. 54.

[13] Armstrong's study of Romantic organicism 'necessarily place[s] a strong emphasis on the functional and philosophical side of organicism rather than its more concretely scientific and biological aspects'. Armstrong, *Romantic Organicism*, p. 5. It seems unlikely, however, that any theory of organicism formulated during the Romantic period failed to at least try to incorporate new insights from the biological sciences.

[14] Neil Vickers, *Coleridge and the Doctors: 1795–1806* (Oxford University Press, 2004), p. 161.

[15] Kevis Goodman, 'Reading Motion: Coleridge's "Free Spirit" and Its Medical Background', *European Romantic Review*, 26.3 (2015), pp. 349–356 (p. 350), doi:10.1080/10509585.2015.1028134. Through his friendships with Thomas Beddoes and Wedgwood, Coleridge was familiar with Brunonian medicine, which explained health and disease in terms of a balance of excitability. He also knew Darwin's medical theories, which hinged on the idea of keeping the organic forces of the body in balance. See 'Coleridge, Thomas Beddoes and Brunonian Medicine', *European Romantic Review*, 8.1 (1997), pp. 47–94.

Whytt's mid-eighteenth-century medical theories, I examine Coleridge's organicism in the joint contexts of vitalism and mind science, which he continued to follow well into the nineteenth century when John Abernethy and William Lawrence had their famous quarrel about the source of the lifeforce. Budge also turns to medical theories, examining the influence of Alexander Crichton's ideas about digestion and the mind in *Inquiry into the Nature and Origin of Mental Derangement* (1798) and Wedgwood's 'organicist account of perception' on Coleridge's theories of language, metaphor, and the symbol.[16] I build on Budge's assertion that the 'focus of his attention is the materially embodied situation in which human thought takes place' and contend that Coleridge's interest in the relationship between the body, world, and mind was driven by his commitment to theorizing an organicism that accounts for relationality among the parts that comprise the universe.[17] Raimonda Modiano reminds us that theorizers of organicism, such as the *Naturphilosophen*, wanted to 'explain how all phenomena of nature were organically related', and Coleridge is no exception.[18]

I rely on Emily B. Stanback's claim that Coleridge's pained body significantly 'influence[d] his conceptions of the human mind, the human imagination, and the divine' as well as his poetry.[19] Where Hartley suffered from painful kidney stones, Coleridge was tormented by a host of physical ailments, including his addiction to opium. His fraught relationship with his body undoubtedly provided a strong impetus to pursue metaphysics and accounts for the increasing abstraction that characterizes the work he published after his return from Malta (where he went to recover from his addiction or to die). Despite this, just as metaphysics do not really provide a release from the body in pain, Coleridge could not completely erase the body from his speculations about vitalism and cognition in 'Theory of Life' (composed 1816) and *Biographia Literaria* (1817). Indeed, late eighteenth- and early nineteenth-century debates surrounding cognition and vitalism centred on the body, and this, in conjunction with Coleridge's extensive medical and scientific knowledge, led to the body necessarily factoring into his theories, even if in sublimated form. In this chapter, I argue that these two texts implicitly assume a mind that is embodied and embedded and illuminate the extent to which the body underpins Coleridge's ideas about knowledge production, organicism, and aesthetics.

Drawing on Perry's method in *Coleridge and the Uses of Division* (1999), I

[16] Budge, *Romanticism, Medicine and the Natural Supernatural*, p. 116.

[17] Budge, *Romanticism, Medicine and the Natural Supernatural*, p. 97.

[18] Raimonda Modiano, *Coleridge and the Concept of Nature* (Florida State University Press, 1985), p. 141.

[19] Emily B. Stanback, *The Wordsworth-Coleridge Circle and the Aesthetics of Disability*, Palgrave Studies in Literature, Science and Medicine (Palgrave Macmillan, 2016), p. 188.

argue that the role of the body in Coleridge's organicism challenges critical commonplaces about Romantic organicism's investment in quasi-Platonic idealism. Despite a long history of scholarship that holds up Coleridge as the apotheosis of transcendental organicism, I contend that his theory belongs in the British tradition of embodied organicism, alongside those formulated by Hartley and Darwin. Perry paints Coleridge as a divided figure who valued 'the immediacy of sense experience', in all its glorious particularity, even as he longed for 'an experience of higher oneness'.[20] These incongruous commitments to what McFarland deems the 'it is' and the 'I am', that is, the external and internal world, drive the duality that Armstrong detects in Coleridge's organicism.[21] Using contemporary 4e cognitive science, phenomenology, and dynamic systems theory as theoretical frames through which to read Coleridge's theories of mind and life, I argue that he attempts to synthesize his phenomenological sense of a 'higher oneness' with nature's self-evident diversity. Like Budge, I move away from ontology, towards lived experience, to argue that the complicated account of the lifeforce in 'Theory of Life', which hypostatizes the abstruse ideas about cognition in *Biographia*, provides insight into how Coleridge tried to theorize his sense of the interconnectedness endemic to lived experience. Additionally, taken together, they provide a basis for interrogating his organicism as an active aesthetic methodology that facilitates human beings' apprehension of divine truth.

In contrast to Fairer, who abandons organicism as a concept because its 'usefulness as a term of critical inquiry would seem to have drained away amid self-contradiction, abstraction, and evasiveness', I interrogate Coleridge's concept of organicism through a 4e lens in order to rescue it from the ashbin of Romantic scholarship.[22] I argue that, as with the other models of organicism discussed in this book, Coleridge offers an important metaphor for structuring the way humans envision themselves in relationship to their surroundings and each other and an aesthetics that encourages lived experiences of relationality. By looking to phenomenology rather than ontology, the apparent 'self-contradiction' becomes moot, the 'abstraction' more embodied, and the 'evasiveness' resolves into a theory that attempts to account for the phenomenological experience of a world that discloses itself as a coherent riot of diversity.

Coleridge's interest in theorizing relationality in the first decades of the nineteenth century drove and even superseded his incursions into ontology. Indeed, in my reading of 'Theory of Life', *Biographia*, and 'Essays on the Principles of Method' published in *The Friend* (1818), Coleridge's investments

[20] Perry, *Coleridge and the Uses of Division*, p. 64.

[21] McFarland, *Coleridge and the Pantheist Tradition*, p. 91.

[22] Fairer, *Organising Poetry*, p. 28.

reflect a commitment to the concept of relationality rather than to a given ontological system.[23] He was initially attracted to Hartley and Priestley because they advanced a theological refutation of Descartes's separation of humankind from its natural environment. Hartley's system not only envisions human beings in relationship with their surroundings, but it also accounts for how this connection is integral to the divine plan, a point that was particularly important to Coleridge. As his later enthusiasm for then rejection of Friedrich Schelling's *System of Transcendental Idealism* (1800) and the names of his two oldest sons (Hartley and Berkeley) indicate, none of the ontological approaches in the mind–matter debates worked for him. Caught between the Scylla of Spinoza and the Charybdis of German idealists, he laboured to account for the fundamental connectedness of the physical, vital, and human orders—to use Maurice Merleau-Ponty's terminology.[24] By framing his speculations in 'Theory of Life' as a hypothesis, Coleridge 'deliberately refuses to credit his claim as ontological'.[25] In fact, I contend that he prioritizes relationality over ontology, the latter of which stems from and subserves his theory of organicism. Thus, in 'Theory of Life' and *Biographia*, Coleridge reframes materiality so that he can connect the body and the physical world to the mental and the divine, irrespective of ontology.

Theories of 4e cognition, particularly their emphasis on embodiment, embeddedness, and enaction, recuperate Coleridge's organicism for a modern audience and establish its relevance in the twenty-first century. Highlighting humankind's situation in a biological environment, the enactive approach combines phenomenology with dynamic systems theory to explain how human beings produce knowledge and to account for how subsystems develop within larger systems. Factoring in human experience and complex systems' unpredictable behaviour, these theories provide a new way to understand Coleridge's attempt to preserve freewill even as he locates cognition in the physical world. In 'Theory of Life', Coleridge tries to envision how organisms acquire form and function in a network that includes the mundane and the divine. He contextualizes human experience 'in terms of the entire sensorium of other living beings and their own autopoietic ways of "bringing forth a world"',

[23] Richard Saumarez's influence, evident in the *Biographia*, likely reinforced the priority Coleridge grants relationships over ontology. According to Richard C. Sha, Saumarez gives preference to 'relationality over difference within physiology' and 'highlights a need to look for and underscore cooperation'. Richard C. Sha, *Imagination and Science in Romanticism* (Johns Hopkins University Press, 2018), p. 148.

[24] For a discussion of Coleridge's lifelong relationship to Spinoza's philosophy, see Richard Berkeley, 'The Providential Wreck: Coleridge and Spinoza's Metaphysics', *European Romantic Review*, 17.4 (2006), pp. 457–475.

[25] Sha, *Imagination and Science*, p. 149.

to borrow a description from posthumanist Cary Wolfe.[26] Aspects of the 4e approach offer a way to account for Coleridge's '*and yet*'.[27] They elucidate how Coleridge's organicism situates humankind within the 'entire sensorium', which Descartes had relegated to the material and therefore mechanical domain, while also preserving its connection to the divine. Explaining the physical world and life as what 4e calls enactive processes, 'Theory of Life' highlights a deep commitment to relationality that serves as the structuring principle of Coleridge's organicism.

Building on phenomenology, which attends to how the world discloses itself to human beings, I turn to neurophenomenology to provide further insight into the relationship between internal and external structures that shape experience and produce knowledge in Coleridge's organicism.[28] As the study of structures of the world and in the body that facilitate perception, neurophenomenology offers insight into embodied aesthetics by providing a model for Romantic structures of engagement, the set of structures that exist in the embodied mind and the world that are responsible for embodied experiences associated with higher states of awareness resulting from aesthetic experiences. They are also the subset of cognitive and environmental affordances that enable experiences often deemed transcendent by the Romantics and the scholars who followed them. I contend that Coleridge's constant 'negotiation between Idea and its embodiment' can be understood as a structural problem, the solution of which is the balance of outward-focused objectivity with inward-focused subjectivity that manifests in human beings. In addition to granting access to so-called objective knowledge about the world, it makes available knowledge of God via Romantic structures of engagement, which facilitate such knowledge through aesthetic encounters with nature and art.[29]

Thus, Romantic structures of engagement and dynamic systems theory reframe Coleridge's organicism as a process, rather than a static ideal, that encompasses relationships that comprise the world and that can be experienced by actively engaging with one's surroundings. For Coleridge, the simultaneous experience of relationality and unity, I contend, is the experience of the

[26] Cary Wolfe, *What Is Posthumanism?*, Posthumanities, 8 (University of Minnesota Press, 2010), p. xxv.

[27] Perry uses the phrase '*and yet*' to describe a pervasive 'syntactic turn' in his writing that 'articulates Coleridge's divided vision' and longing to have it both ways. Perry, *Coleridge and the Uses of Division*, pp. 23–24.

[28] Originating with Charles D. Laughlin, John McManus, and Eugene G. D'Aquili, *Brain, Symbol & Experience: Towards a Neurophenomenology of Human Consciousness* (Shambala, 1990), neurophenomenology characterizes early work by Francisco Varela and Evan Thompson, who, although they did not come up with the term, laid the foundations of 4e theories of cognition along with Eleanor Rosch.

[29] Fairer, *Organising Poetry*, p. 9.

divine. Not only does it rely on the body, it also must be actively sought by interacting with nature, poetry, and philosophy. Hence, his aesthetic method takes advantage of Romantic structures of engagement to 'surprise and shock us, to create conflict, and to lure the [readers] into unaccustomed ways of [reading] that can allow meaningful, self-motivated emotional processes to unfold'.[30] His poetry offers opportunities to 'tremble into thought' as readers experience an embodied sense that they belong to a web of relations made up of infinite multeities and draw their own conclusions about the nature of life, being, and God.[31] Focusing on two texts that rely on the concept of embodiment even as they obfuscate the physiological body, I argue that the model of organicism that Coleridge fleshes out in 'Theory of Life' privileges relationality over ontology and that it provides a frame for understanding the rarefied epistemology in *Biographia* as a theory of embodied cognition that underwrites his organicism. I end the chapter with an analysis of 'The Eolian Harp' that demonstrates how his aesthetics offer a method to discipline the mind and body into active habits of engagement with and awareness of the vast web of relations that comprises the internal and external world in order to access fundamental principles of life.

Descartes and the Problem of Ontology

In *Discourse on Method* (1637), Descartes links epistemology with ontology, thus introducing a 'disabling empiricist gap between idea and material reality' that occupied philosophers for centuries.[32] While some solutions to the problem of mind–body–world interaction strove to preserve the mind–matter dichotomy, Louis de la Forge's seventeenth-century theory of occasionalism for example, others, such as the one advanced by Anthony Collins, who suggested that cognition is an emergent property of the 'combined activities of the particles of the brain', rejected Cartesian ontology outright.[33] As physics took a Newtonian

[30] Ralph D. Ellis, *Curious Emotions: Roots of Consciousness in Motivated Action* (John Benjamins Publishing, 2005), p. 170.

[31] Coleridge, 'Eolian Harp', l. 46.

[32] Perry, *Coleridge and the Uses of Division*, p. 111.

[33] Liam P. Dempsey, 'Thinking-Matter Then and Now: The Evolution of Mind-Body Dualism', *History of Philosophy Quarterly*, 26.1 (2009), pp. 43–61 (p. 50). One of Descartes's first 'self-styled disciples', La Forge advanced the idea that neither bodies nor human minds can act on matter and that only God can cause movement in bodies. Andrea Sangiacomo and Desmond Clarke, 'Louis de La Forge', in *The Stanford Encyclopedia of Philosophy*, ed. by Edward N. Zalta and Uri Nodelman, Summer 2020 (Metaphysics Research Lab, Stanford University) <https://plato.stanford.edu/archives/sum2020/entries/la-forge/> [accessed 1 June 2022]. Collins proposed this theory as part of an extended exchange with committed dualist Samuel Clarke in the late 1730s; see Dempsey, 'Thinking-Matter Then and Now'.

turn and the body increasingly became an object of scientific curiosity, it became less viable to think of either nature or the body as mechanical. Coleridge blamed Descartes for introducing the possibility of mechanistic accounts of life, naming him 'the first man who made a direct division between man and nature, the first man who made nature utterly lifeless and Godless, considered it as the subject of merely mechanical laws'.[34] Not only did Descartes evacuate nature of its very essence—activity, vitality, life—he also committed phenomenological and theological solecisms by making assertions contrary to most people's lived experience of nature and by severing the connection between God and his creation. Furthermore, finding solutions for the host of problems raised by Cartesian epistemology became more untenable as knowledge of physiology increased.

Cartesianism, or the 'generalizing philosophy' as Coleridge calls it in a note on Baruch Spinoza, 'had gone on refining and refining on the ~~nature~~ spiritual nature of God' until it had become 'a mere Abstraction to the mind, virtually excluding the divine presence from the creation … leaving the idea of God Natureless, and Nature Godless'.[35] Since Cartesianism dominated general philosophy well into the nineteenth century, one of Coleridge's philosophical concerns was to repair the schism it had imposed on nature, separating God from the physical world and alienating people from their intuitive sense that nature is vibrant, alive, and (sometimes) awe-inspiring. He does this by reframing the physical world in terms of relationality. Trying to restore to nature what Descartes had 'emptied' from it—'life and all that makes it nature in reality'—Coleridge's first task involved finding a way to incorporate self-propelled activity into the natural world while keeping God connected yet discrete.[36] The second was to understand how the mind communicated with the world via the body and vice versa, without falling into necessitarianism. The first question drove the vitalist movement, while the second motivated the mind–matter debates. Where Hartley focuses on cognition and Darwin emphasizes life, Coleridge pursues both questions equally. In 'Theory of Life', he theorizes how disparate forms of life exist in a physical substrate that maintains itself as a stable, unified system, while *Biographia* comes at the

[34] Samuel Taylor Coleridge, *Lectures 1818–1819: On the History of Philosophy*, ed. by J.R. de J. Jackson, The Collected Works of Samuel Taylor Coleridge, 8, 2 vols (Princeton University Press, 2000), ii, p. 256.

[35] Samuel Taylor Coleridge, 'Note on Spinoza', *The Collected Works of Samuel Taylor Coleridge: Shorter Works and Fragments*, ed. by H. J. Jackson and J. R. de J. Jackson, Collected Works of Samuel Taylor Coleridge, 11, 2 vols (Princeton University Press, 1995), i, p. 623.

[36] Coleridge, *Lectures*, ii, p. 565. Sha's discussion of Coleridge's reframing of the subject and object as modes of knowing is useful here. See Sha, *Imagination and Science*, pp. 175–176.

problem from the perspective of cognition, probing the nature of the relationships that enable living beings to perceive and understand the world in which they live. Thus, Coleridge's investigations into vitalism, consciousness, and freewill were driven by larger questions about relationality. He did not merely ask whence life came but sought to understand 'what makes order possible and how highly organized complexity, which is highly improbable, comes into being at all', a question posed by proponents of twenty-first-century systems theory, whose work provides insight into Coleridge's ideas.[37]

Eighteenth- and nineteenth-century responses to this twofold problem—how to account for vitality in nature and for mind–body–world interactions—were often couched in ontological terms. While Coleridge did think about ontology, as his worries about the pantheistic implications of materialism and idealism indicate, 'Theory of Life' does not consider the question on ontological grounds, and reading it that way has led critics to inaccurately characterize Coleridge as a dualist. Such readings miss the stakes of his organicism—providing a viable alternative to Cartesianism that accounts for his phenomenological intuition that the universe is best characterized by the principle of relationality, '*and yet*' it is also best characterized by the principle of unity.[38] Human beings experience the physical world as both. In 'Theory of Life', Coleridge notes that 'visible SURFACE and *power* of any kind, much more the *power* of life, are ideas which the very forms of human understanding make it impossible to identify'.[39] In a Kantian vein, he asserts the unfeasibility of certain knowledge about the so-called essence of objects.[40] His reference to the 'forms of human understanding' implies that human perception constrains certain knowledge, and he gestures towards the phenomenological nature of knowledge.

Looking at 'Theory of Life' from the perspective of phenomenology provides a better understanding of how Coleridge conceives relationality irrespective of ontology, and the text invites such a reading. He makes it clear that he is not 'submitting [his] own ideas on the subject of life' but 'rather [on the form] in which it is manifested to *us*'.[41] He attends to the phenomena that are available to the senses. Acknowledging that life cannot be perceived directly but only by its effects, Coleridge attempts to induce the common 'characteristic instinct or tendency, evident in all its manifestations and involved in the idea itself'.[42] His use of the terms 'manifestation' and 'idea' suggests a phenomenological focus

[37] Wolfe, *What Is Posthumanism?*, p. 13.

[38] Perry, *Coleridge and the Uses of Division*, p. 24.

[39] Coleridge, 'Theory of Life', *Shorter Works*, I, p. 503.

[40] Sha argues that Coleridge frames 'Theory of Life' as a hypothesis, which implies Coleridge's ontological modesty, a point with which I agree. See Sha, *Imagination and Science*, pp. 146–150 and 165–171.

[41] Coleridge, 'Theory of Life', *Shorter Works*, I, p. 500.

[42] Coleridge, 'Theory of Life', *Shorter Works*, I, p. 504.

'not on *what things are* but on the *ways in which they are given*', that is, the way the world discloses itself to the percipient.[43] As a physical phenomenon that can be perceived by the human mind but that is also 'involved in the idea' of life 'itself', its 'characteristic tendency' points to the link between percept and fact.[44] Coleridge sees it as a single phenomenon that exists both as a *ding-an-sich* and as a thing that can be known by human beings.

Coleridge's 'Essays on the Principles of Method', in which he develops the concept of phenomenology, further suggest that 'Theory of Life' should be read through such a lens. His method indicates that human beings can know divine truth, which he calls 'LAW', by cultivating the faculty of reason.[45] However, his discussion of the second aspect of method, 'THEORY', acknowledges that humans come to this knowledge not just by observing nature but, more importantly, by experiencing it. By attending to 'the existing forms and qualities of objects, discovered by observation or experiment, [which] suggest a given arrangement of many under one point of view', people come to recognize that relationality structures the physical world.[46] As embodied creatures, human beings have access to nature only in its 'material sense' or, as Coleridge elaborates, 'the sum total of all things, as far as they are objects of our senses, and consequently of possible experience'.[47] He calls this observable aspect of nature 'Phænomenology', a study that encompasses both embodied and perceptual experience, which he subdivides into 'Somatology and Psychology'.[48] He suggests that by closely observing natural phenomena as well as one's experience of it, people can come to consciously know the intuitive principle of simultaneous unity and multeity.

Coleridge's description of life in 'Theory of Life' as both manifestation and cause recalls Johann Friedrich Blumenbach's account of the lifeforce as a 'power' in which the formative 'nisus stems from its ability to be both cause and effect'.[49] According to Sha, Coleridge 'looked for indirect evidence of [life] that would confirm it as a kind of causal law'.[50] Coleridge's stated goal in 'Theory of Life' to 'render the phenomena intelligible as its different degrees

[43] Evan Thompson, *Mind in Life: Biology, Phenomenology, and the Science of Mind* (Harvard University Press, 2007), p. 21.

[44] Coleridge, 'Theory of Life', *Shorter Works*, I, p. 545.

[45] Coleridge uses the terms God, the Law, and the absolute interchangeably, so I do as well. In Coleridge's lexicon, the law and the absolute are, ultimately, God.

[46] Samuel Taylor Coleridge, 'Essay VI', *The Friend*, ed. by Barbara E. Rooke, The Collected Works of Samuel Taylor Coleridge, 4, 2 vols (Princeton University Press, 1969), I, p. 464.

[47] Coleridge, 'Essay VI', *Friend*, I, p. 467.

[48] Coleridge, 'Essay VI', *Friend*, I, p. 467.

[49] Sha, *Imagination and Science*, p. 167. See pp. 165–167, for a discussion of the distinction between effects that can be observed and therefore known, and causes, about which thinkers such as Kant, Blumenbach, and Coleridge were circumspect.

[50] Sha, *Imagination and Science*, p. 146.

and modifications' indicates an attempt to explain how life discloses itself in its varying forms, from the simplest to the most complex organisms, thus making 'Law' recognizable.[51] Bracketing ontology and focusing on phenomenology when investigating Coleridge's model of organicism provides insight, therefore, into his ideas about life and cognition as expressions of organic relationality that attempt to restore what Descartes had torn apart.

Phenomenological Organicism

Written in the second decade of the nineteenth century, when the question of life was at the forefront of philosophical, scientific, and aesthetic discourse, 'Theory of Life' offers a concentrated account of Coleridge's concept of organicism. Originally, it began as a collaborative effort to assist James Gillman, the physician who helped Coleridge get his opium use to a manageable level, win the 1816 Royal College of Surgeons' annual Jacksonian Prize competition. Gillman was to write about contemporary theories and treatments of scrofula, while Coleridge was charged with writing the history of treatment, beliefs, and ideas about it.[52] The history soon ballooned into a full-blown treatise on life. In 'connecting life and diseases, Coleridge follows others, including Erasmus Darwin and the Hunterians', including Abernethy, who espoused the tenets of comparative anatomist John Hunter.[53] Like Darwin, Coleridge believed that people's health was connected to the embodied lifeforce. If it lacked equilibrium, if there was a deficit or surplus of sensibility, an individual would manifest disease.[54] At the same time, thinkers hypothesized that the 'life of an organism, like the aesthetic power of a work of art, cannot be reduced to a component quality or part'.[55] The burgeoning life sciences, notably biology and physiology, helped create the conditions for the public controversy between Abernethy and Lawrence that occurred around the time Coleridge wrote 'Theory of Life'. The two physicians delivered their competing positions as part of the Hunterian lecture series at the Royal College and

[51] Coleridge, 'Theory of Life', *Shorter Works*, I, p. 505.

[52] Scrofula refers to a disease characterized by swollen lymphatic glands; related to tuberculous, it was also known as the King's Evil.

[53] Tilottama Rajan, 'The Unavowable Community of Idealism: Coleridge and the Life Sciences', *European Romantic Review*, 14.4 (2003), pp. 395–416 (p. 408).

[54] Vickers traces the evolution of Coleridge's beliefs about disease from those espoused by Darwin (materialist) to those advanced by Beddoes (mentalist). The materialist hypothesis attributed disease to imbalances with physical causes, such as 'an overactive tactile sense'. Vickers, *Coleridge and the Doctors*, p. 163. Beddoes's theory, on the other hand, looked to 'long-entrenched mental dispositions' as the cause of physiological disorders. Vickers, p. 162.

[55] Denise Gigante, *Life: Organic Form and Romanticism* (Yale University Press, 2009), p. 23.

later published them, thus provoking Coleridge's response.[56] Unsurprisingly, Coleridge and Gillman did not finish the essay on scrofula in time to submit it to the Jacksonian competition; surprisingly, however, it was Gillman and not Coleridge who caused the delay.

Despite finishing 'Theory of Life', Coleridge did not publish it, and it did not receive scholarly attention until it was discovered posthumously among Gillman's papers.[57] Since then, it has been held up by literary critics as a paradigmatic example of transcendentalist idealism. However, I argue that the model of organicism that Coleridge advances in 'Theory of Life' is not a 'purely conceptual unity achieved at the expense of experience' but an attempt to account for his twofold phenomenological sense that the physical world comprises a temporal succession of discrete entities and simultaneously exists as a unified system that seemingly defies time.[58] Furthermore, Coleridge's interpretation of his experience is consistent with Hartley's notions that humankind grows in understanding of and closeness to God by studying nature. Thus, Coleridge theorizes organicism based on his lived experience that nature is not 'the subject of merely mechanical laws'.[59] He, and many others, experience it as an interconnected set of vital phenomena that provide the conditions of possibility for creativity, communion with God, and, paradoxically, life itself.

Sharing Priestley's motivation for reconceptualizing matter in *Disquisitions Relating to Matter and Spirit* (1777), Coleridge sought to controvert Cartesianism by reconnecting the body with the mind and the world with the divine. However, in contrast to Priestley, he rejected the determinism entailed by materialist accounts. Yet, even as he thinks through an alternative to the

[56] Abernethy delivered his lectures on 'Mr. Hunter's Theory of Life' in 1814 and 1815, and Lawrence gave his lecture, titled 'An Introduction to Comparative Anatomy and Physiology' in 1816. These were followed by the publication of Abernethy's *Introductory Lectures, exhibiting Some of Mr. Hunter's Opinion respecting Life and Diseases, delivered before the Royal College of Surgeons in 1815* and Lawrence's *Lectures on Physiology, Zoology, and the Natural History of Man: Delivered at the Royal College of Surgeons in 1819*. See Sharon Ruston, '"Natural Enemies in Science, as Well as in Politics": Romanticism and Scientific Conflict', *Romanticism*, 11.1 (2005), pp. 70–83, especially p.71.

[57] For information about the composition and publication history of 'Theory of Life', see Trevor H. Levere, *Poetry Realized in Nature: Samuel Taylor Coleridge and Early Nineteenth-Century Science* (Cambridge University Press, 1981), pp. 215–219; Robert Mitchell, *Experimental Life: Vitalism in Romantic Science and Literature* (Johns Hopkins University Press, 2013), Chapter 3; Alice Dorothea Snyder, *Coleridge on Logic and Learning* (New Haven, 1929), HathiTrust Digital Library <http://hdl.handle.net/2027/mdp.39015074764096> [accessed 1 June 2023]; see Jackson and Jackson, Introductory Notes to 'An Essay on Scrofula', in *Shorter Works*, I, by Samuel Taylor Coleridge, pp. 454–456, and Introductory Notes to 'Theory of Life', in *Shorter Works*, I, pp. 481–484.

[58] Jerome McGann, *The Romantic Ideology: A Critical Investigation* (University of Chicago Press, 1983), p. 38.

[59] Coleridge, *Lectures*, II, p. 565.

materialist theories of life advanced by Lawrence, Thelwall, and the French *philosophes*, Coleridge does not advance a 'Plotinian theory of evolution' that borrows from 'contemporary scientific theory'.[60] Neither is it accurate to claim that 'Coleridge was sufficiently a Platonist, better a Plotinist', or that the 'powers and forces' he describes 'are not physical or phenomenal' but transcendental powers that 'cannot be pictured, but only imagined'.[61] Coleridge does not propose a metaphysics to ground his organicism. Rather, his theory of how the physical world gives rise to interdependent but discrete living entities, many with high-level cognitive capabilities, privileges the principle of relationality.

In 'Theory of Life', he proposes a relationality so deep that he weaves it into the structure of the universe even before life makes its first appearance on Earth. Like Darwin in *The Temple of Nature* (1803), Coleridge offers an account of how the tangible universe comes into being, wherein he establishes the milieu in which life is embedded and the process by which it manifests in living bodies. In this discussion, he represents the structure of the phenomenal world as similar to the manifestation of the lifeforce. Just as living, cognizing bodies result from the striving of the two polar ends of life (unity and multeity) that manifest in increasingly complex species, the inanimate universe emerges from two primary building blocks, which Coleridge conceives as aspects of God. These two originary elements are space and time.

Coleridge's assumptions about space and time reflect a post-Cartesian conception that allows him to treat them as phenomena that are implicated in and as mysterious as God himself. In *Principles of Philosophy* (1644), Descartes 'lays out the thesis of the identity of space (extension) and matter', thus conceptualizing space as a material substance.[62] However, Newtonian physics revolutionized how thinkers understood these two phenomena. They did 'not regard space and time as genuine substances (as are, paradigmatically, bodies and minds), but rather as real entities with their own manner of existence as necessitated by God's existence (more specifically, his omnipresence and eternality)'.[63] The affiliation of space and time with divine mysteries worked well for Coleridge, who took it for granted that God is 'the ground or cause' of everything.[64] In fact, he did not consider space entirely real, but neither

[60] Vallins, *Coleridge and the Psychology of Romanticism*, p. 135.

[61] C.U.M. Smith, 'Coleridge's "Theory of Life"', *Journal of the History of Biology*, 32.1 (1999), pp. 31–50 (pp. 38–39); M.H. Abrams, *The Correspondent Breeze: Essays on English Romanticism* (W.W. Norton, 1984), p. 171.

[62] Robert Rynasiewicz, 'Newton's Views on Space, Time, and Motion', *The Stanford Encyclopedia of Philosophy*, ed. by Edward N. Zalta and Uri Nodelman, Summer 2014 (Metaphysics Research Lab, Stanford University) <http://plato.stanford.edu/archives/sum2014/entries/newton-stm/> [accessed 25 November 2016].

[63] Rynasiewicz, 'Newton, Space, Time, Motion'.

[64] Coleridge, 'Theory of Life', *Shorter Works*, I, p. 503.

did he view it as ideal. He asserts that to 'make it wholly ideal, would be at the same time to idealize all phenomena', an ontological move that he rejects because it would 'undermine the very conception of an external world'.[65] Yet to 'make it real' is equally absurd; thus he concludes that '*space* is the ideal organ by which the soul of man perceives the *omnipresence* of the Supreme Reality'.[66] Space, in Coleridge's view, is one of two elements from which the phenomenal world emerges and makes the 'LAW' intelligible to the human mind.

Coleridge hypothesizes that, parallel to the polarity that characterizes the law of life, space and time represent the two poles of God's mysteries that productively interact to form the world. He links space with '*omnipresence*', which he claims distinguishes God 'from the works, which in him move, live and have their being'.[67] In other words, space corresponds to and authorizes the principle of individuation. This is the mystery of 'multeity' for which Coleridge tries to account in his model of organicism. Time, on the other hand, coincides with 'his *Eternity*, or what', according to Coleridge, 'is fully equivalent, his Unity'.[68] Thus, these two elements correlate with the twofold mystery that Coleridge attempts to resolve in his theory of organicism. As space and time enter into relationship with each other, they produce new elements or forces that become available to enter into new relationships, and in this process the physical world emerges and gradually becomes perceptible.

For Coleridge, the initial intersection of time and space produces motion, which once it exists it enters into a relationship with time and space, thus producing line. Next, line intersects with time, space, and motion to produce new powers, and on and on it goes until all the Newtonian forces have been proliferated. When the 'simple ideas of time, space, and motion; of length, breadth, and depth' enter into relationships with each other, they create Coleridge's infamous '*tertium aliud*', or the 'third other thing'.[69] This ongoing

[65] Coleridge, 'Theory of Life', *Shorter Works*, I, p. 520.

[66] Coleridge, 'Theory of Life', *Shorter Works*, I, p. 520.

[67] Coleridge, 'Theory of Life', *Shorter Works*, I, p. 520. Sha discusses the importance of hypothesis and postulation to Coleridge's method of arriving at his theory of imagination, explaining that 'hypothesis needed to be testable either by logic or experimentation or supported by facts or laws'. Sha, *Imagination and Science*, p. 145. Coleridge, he asserts, rejected untenable hypotheses that did not take embodied experience into account. Postulates, on the other hand, are necessary to enable him, or any philosopher, to proceed in theory-building. A postulate asserts the truth of a given axiom and brackets questions of its facticity or ontological status. It is important to bear in mind that Coleridge acknowledges the provisional status of his theory, admitting that he presents it 'not as a truth … but as an hypothesis, which pretends no higher merit than that of explaining the particular class of phenomena to which it is applied'. Coleridge, 'Theory of Life', *Shorter Works*, I, p. 517.

[68] Coleridge, 'Theory of Life', *Shorter Works*, I, p. 520.

[69] Coleridge, 'Theory of Life', *Shorter Works*, I, p. 522; Coleridge, 'Theory of Life', *Shorter*

process of interacting and entering into relationships not only produces all the nontangible phenomena, but eventually the tangible ushers forth from it. This recursive process of joining and re-joining, Coleridge contends, continues 'even till they [the products] pass into outward phenomena'.[70] He calls this process of structuration 'interpenetration' or 'total intussusception'.[71] A relatively new word in the nineteenth century, the noun intussusception hypostatizes the process by which one thing absorbs another. In physiology and biology, it refers to the assimilation of new materials into a living organism that are then converted into organic tissue. Coleridge's use of the term, which he presumably learned from his medical and scientific reading, is similar. As a new force is produced, it assimilates its comprising forces and converts them into something with a new, but related, set of properties. Then, it disperses itself among the extant forces to repeat the process.

Coleridge provides equations to make visible his ideas about the proliferation of Newtonian forces from relationships of time and space and the forces that they produce. His notational representation of intussusception, in conjunction with his verbal explanation of space and time, brings the fundamental relationality of the universe sharply into focus and illustrates the importance of structuration to his organicism:

> Time + space = motion;
> T*m* + space = line + breadth = depth;
> depth + motion = force;
> L*f* + B*f* = D*f*;
> LD*f* + BD*f* = attraction + repulsion = gravitation.[72]

Despite their seeming nonsense, the syntax and semantics in this series of equations offer crucial information for interpreting Coleridge's organicism.

Demonstrating the logic of intussusception, Coleridge uses variables (L for line, *f* for force, and so forth) to signify that time and space are not only discrete phenomena but so are all the forces that they produce. Each force exists and functions differently from the others. To illustrate how they enter into productive relationships with each other, he connects them with mathematical symbols. The plus signs indicate combinational but not merely additive relationships between forces. The expression 'Time + space = motion' represents a process by which time combines with space to produce motion,

Works, I, n. 1, p. 522.

[70] Coleridge, 'Theory of Life', *Shorter Works*, I, p. 522.

[71] Coleridge, 'Theory of Life', *Shorter Works*, I, p. 525.

[72] Coleridge, 'Theory of Life', *Shorter Works*, I, p. 522. Coleridge integrates these symbols into the prose of his paragraph, which makes them difficult to read and easy to overlook. For that reason, I put each equation on a separate line.

but rather than simple accretion the addition of time to space is more akin to the joining of hydrogen and oxygen molecules to produce water. The original elements are not simply increased; they join to yield an entirely different force, motion. The next expression, '*Tm* + space = line', shows that time (*T*) and space have retained their original properties and continue to exist as they were prior to entering into relationship with each other. Motion (*m*) represents the new intussuscepted force that has assimilated and converted properties of time and space. Once it exists, motion is available to enter into relationships with the other forces, thus producing line. Yet the individual variables continue to suggest multeity even as their conjunction produces unity.[73]

Using different kinds of notations to represent different types of combinations, Coleridge attempts to represent a system that is in a perpetual state of activity. While impossible to trace ontologically, the combination of 'nonadditive relationships among variables, particularly exponential and multiplicative interactions', when viewed through the lens of dynamic systems theory, suggests the principle of process.[74] While he joins some of the variables with plus signs, several variables stand adjacent to each other in groups of two or three. Lacking an explicit mathematical symbol to link them, their relationships must be deduced by the logic of algebra in which 'multiplication' has long been 'indicate[d] by simple apposition: thus, *ab* signifies *a* multiplied by *b*'.[75] Similarly, *Tm* signifies time multiplied by motion. The product of these two forces is then added to space to produce line, which is then added to breadth to produce depth. Similar combinations of linear and nonlinear operations are used by systems theorists in the twenty-first century to represent systems that are in motion.

Even though dynamic systems theory postdates Coleridge, it helps to make legible his 'doctrine concerning energetic nature [that] is comprised in the science of DYNAMICS' as illustrated by this set of notations in 'Theory of Life'.[76] According to Trevor Levere, in contrast to mechanism, dynamic

[73] A discussion of Coleridge's use of equal signs is not necessary to my argument here. However, for a detailed analysis of them in connection with the rest of the mathematical symbols in these 'equations', see Lisa Ann Robertson, '"Time + Space": Coleridge's Phenomenological Organicism', *Coleridge Bulletin*, n. s., 62 (2024), pp. 3–19.

[74] Luis H. Favela, 'Dynamical Systems Theory in Cognitive Science and Neuroscience', *Philosophy Compass*, 15.8 (2020), pp. 1–5 (p. 3), doi:10.1111/phc3.12695.

[75] Florian Cajori, *A History of Mathematical Notations* (Dover Publications, 1993), p. 224. Cajori translates the above quotation from an article by Gottfried Wilhelm Leibniz published in *Miscellanea Berolinensia* (1710), a journal of the Royal Prussian Society of Sciences. While mathematical notation was contested during the eighteenth and into the nineteenth century, 'mere juxtaposition [had] signified multiplication' in manuscripts 'since the fifteenth century'. Cajori, *A History of Mathematical Notations*, p. 250.

[76] Coleridge, 'Essay VI', *Friend*, I, p. 467.

science meant 'showing nature in all its complex unity and flux'.[77] Even if Coleridge's equations would be considered nonsense in contemporary scientific terms, much as differential equations might seem like gibberish to non-specialists, they are an expression of one of the fundamental principles that undergirds his organicism—perpetual motion that constantly changes states. Systems theory and dynamic systems theory, in particular, offer an instructive parallel. Where the former captures the movement of predictable systems, such as the pendulum on a clock, the latter accounts for those that have irregular or singular outcomes, that is, dynamic systems. The distinction that systems theory makes between determined and dynamic systems helps elucidate Coleridge's ideas. Determined systems are akin to 'mechanism, *ab extra*', which Coleridge explains as 'whatever is organized from without', or by forces outside the system itself; this 'is a product of mechanism'.[78] This kind of phenomenon's future behaviour can be predicted from its past states and can be expressed in straightforward mathematical terms by systems theorists. Like the swinging pendulum or celestial bodies, all its possible trajectories can be mapped. Dynamic systems, on the other hand, have the capacity to develop along so many different trajectories that to express future states requires complex mathematical calculations—differential equations—that capture every possible iteration as they move through time.[79] As their similar names suggest, dynamic systems theory and dynamism recognize the unpredictable behaviour of vastly complex systems.

It is notable that in 'Essays on the Principles of Method', Coleridge asserts that 'the union' of 'Dynamics' with 'Phænomenology' combined with 'Logic and Mathematics, constitute' the proper science of 'Natural Philosophy' because he takes into account both the experience of the phenomenal world and the laws that govern it.[80] The world discloses itself as both unpredictably surprising and obviously regulated in some ways. Dynamics systems theory also tries to account for this curiosity, contending that even though their constant change makes them impossible to predict, these systems nonetheless exist in a deterministic state space.[81] The state space refers to the range of possible futures, and, although unpredictable, they are still limited by a host of circumstances. Similarly, Coleridge's acknowledgement of Newtonian physics indicates that he recognized that the universe—and even life itself—is subject to certain immutable laws, such as gravity. Despite his commitment to the human will,

[77] Levere, *Poetry Realized in Nature*, p. 79.

[78] Coleridge, 'Theory of Life', *Shorter Works*, i, p. 511.

[79] Every second (or even nanosecond) that dynamic systems creep forward in time, they present another set of nearly infinite possible states, making them impossible to predict, at least at represent.

[80] Coleridge, 'Essay VI', *Friend*, i, p. 467.

[81] Favela, 'Dynamical Systems Theory', p. 4.

Coleridge rejected hypotheses, theories, and 'notions that either defy laws of embodiment or that exceed our ability to have some experience of them'.[82] Thus, even though he could not accept the idea of a predetermined universe, he did not embrace a chaotic one in which anything goes. Despite its subjection to regular laws, the physical world—both animate and inanimate—manifests spontaneous activity that leaves room for unexpected developments.

With respect to Coleridge's concept of organicism, dynamism circumvents materialist determinism without foreclosing systems' implication in physicality—even the human system. Driven by many factors, from his desperation to gain control over his fraught embodiment to his religious beliefs, Coleridge's commitment to freewill, in part, corresponds with the phenomenological sense that people have the capacity to make decisions that affect the immediate and long-term outcomes of their lives. Hartley, too, acknowledged this sense of freedom by advancing the notion of constrained choice, or what he calls practical freewill, which contributes to both the feeling and the reality that neither the fate of the world nor any individual being can be definitively predicted. The concept of freewill can be understood in terms of the local instabilities that make it impossible to predict exactly how any dynamic system will unfold over time, despite the theoretical possibility of using a differential equation to plot all the potential state spaces. Local instabilities occur as a system moves through time and space, encountering various circumstances that disrupt its trajectory and to which it must respond. In every encounter, according to Niklas Luhmann, there is a difference 'between what is *actually given*'—the current system state—'and what can *possibly* result from it'—its future state.[83] For every possibility realized there remain uncountable unrealized states that are foreclosed as a system moves forward in time. Enactivists describe this phenomenon in living organisms in terms of perturbations, or environmental stimuli. When an organism encounters stimuli that it deems relevant, it 'will lead to a change in state configuration'.[84] While some patterns emerge, complex and dynamic systems, like the physical world that Coleridge describes verbally and with notations in 'Theory of Life', cannot be adequately explained by the laws of mechanism. Analysing Coleridge's organicism in terms of dynamic systems theory helps resolve the seeming tension between materialist determinism and idealist possibility.

Dynamic systems theory also helps make sense of the paradox that Coleridge's model of organicism seems to entail by making room for self-referentiality. In one of the essays on method, Coleridge asserts that 'the very impulse to universalize in any phænomenon involves the prior assumption of

[82] Sha, *Imagination and Science*, p. 157.

[83] Niklas Luhmann, *Social Systems*, trans. by John Bednarz, Jr. with Dirk Baecker (Stanford University Press, 1996), p. 74.

[84] Varela, Thompson, and Rosch, *The Embodied Mind*, p. 152.

some efficient laws in nature, which in a thousand different forms is evermore one and the same; entire in each, yet comprehending in all; ... it is itself presupposed in each and all as their common ground and condition'.[85] In positing a single principle that serves as the common ground for all phenomena, Coleridge apparently falls into two errors. First, as Armstrong notes, he seems to prioritize the 'impulse to universalize' at the expense of reciprocity between the organizing principle and the organized parts. Second, as Jerome Christensen contends, even though Coleridge 'conceives of himself and everything else as necessarily within the system', he must 'appeal to an authoritative voice', God, or a common ground that exists '*ab extra*'.[86] The problem noted by both critics is that Coleridge's self-contained system relies on something external to it, which undermines the integrity of a model of organicism predicated on the integration of parts and whole. Considered from the perspective of dynamic systems theory, a system differentiates itself from the surround from which it takes it origin by creating boundaries and processes that distinguish it from the larger system in which it exists. The paradox is a self-contained system that takes its origin from and subsists in another system.

For Coleridge, God provides the original conditions of possibility. Because eternity and omnipresence give rise to time and space and, therefore, the physical world, the latter exists in relationship to and within God, who enables its existence to begin with. This seeming 'paradox', claims Luhmann 'does not prevent the operations of the system', but 'on the contrary, it is the condition of their possibility'.[87] Systems serve as environments for subsystems, which define their difference in terms of systemic unity. Luhmann borrows the idea of autopoiesis, by which living systems organize themselves from within and distinguish themselves from their surrounds, from enactive biologists Humberto Maturana and Francisco Varela and applies it to non-living systems (despite Varela's objections). According to autopoiesis, organisms emerge from an environment by forming cell boundaries that keep them intact and distinct from their milieu. These boundaries protect the mechanisms that execute the metabolic processes that sustain an organism's life. Considering a cell or an organism, it is easy to envision a self-contained entity that emerged from and exists in the system against which it defines itself. The paradox of simultaneous identity and non-identity of a system that exists within a system, rather than 'hindering the autopoiesis of self-referential systems, in fact *force*[s] their autopoiesis'.[88] Autopoiesis helps account for systems that are autonomous from *and* dependent on their milieu.

85 Coleridge, 'Essay VI', *Friend*, I, p. 467.

86 Christensen, *Coleridge's Blessed Machine*, p. 57.

87 Niklas Luhmann, *Theories of Distinction: Redescribing the Descriptions of Modernity*, trans. by Joseph O'Neil, Elliot Schrieber, Kerstin Behnke, and William Whobrey (Stanford University Press, 2002), p. 84.

88 Cary Wolfe, 'Meaning as Event-Machine, or Systems Theory and "The Reconstruction

The autopoietic process sheds light on Coleridge's concept of organicism, particularly as depicted in the construction of the physical world in 'Theory of Life'. The paradoxical relationship between God (unity) and the physical world (multeity) can be understood as a dynamic system. The formation and maintenance of such a system is a process or ongoing act of differentiating itself from its milieu. Far from being a paralyzing contradiction, the productive tension between the multeity and unity, or closure and openness, keeps the system intact. The constant drive to maintain itself—think of Kant's purposiveness without purpose or Spinoza's conatus—propels the system forward in time and space. Without this tension, the system would dissolve into its environment and become part of the undifferentiated mass, that is, it would be absorbed into unity, which is precisely how Coleridge describes death. If, as Coleridge's account suggests, phenomenality is based on relationality as forces enter into relationships that make the world tangible, it is but a short step to the self-organizing energy that Coleridge calls the 'vital principle'.[89] Rather than presenting an inimical obstacle to overcome, physicality provides the conditions for both life and sentience to obtain.

Dynamic systems theory reveals that in 'Theory of Life', irrespective of ontology or theology, Coleridge tackles the problem of how to account for continuous temporality in systems that have 'processes that are not discrete' from the universalizing principle but nonetheless individuate into discrete phenomena.[90] The physical world, like life, persists even as individual objects and organisms come and go. Indeed, the existence of 'the "Non-Absolute" is the primal differentiation from God that makes reality a realm of dynamic becoming, as opposed to timeless stasis'.[91] When read through the lens of dynamic systems theory, then, Coleridge's notations show that he envisions even the non-living aspects of the world as an ongoing and active process. The relationality that comprises every aspect of the phenomenal world is the fundamental principle that drives Coleridge's theory of organicism.

The connection between systems and their environment entails relationality not just because the system requires its environment to survive but also because it requires it to form itself in the first place. Systems can differentiate themselves, but they cannot avoid being in relationship. In Coleridgean terms, the forces that emerge from time and space create a self-contained system, but not one that is disconnected from God. The relationship between the physical

of Deconstruction": Derrida and Luhmann', in *Emergence and Embodiment: New Essays on Second-Order Systems Theory*, ed. by Bruce Clarke and Mark B. N. Hansen, Science and Cultural Theory (Duke University Press, 2009), pp. 220–245 (p. 230).

89 Coleridge, 'Theory of Life', *Shorter Works*, I, p. 532.

90 Favela, 'Dynamical Systems Theory', p. 7.

91 Gregory Leadbetter, *Coleridge and the Daemonic Imagination* (Palgrave Macmillan, 2011), p. 49.

world and God is primary; it creates the conditions for all others to come into being. Coleridge's account of how objects become perceptible implies an embodied subject. Although highly abstract, the account of life and cognition in 'Theory of Life' and *Biographia Literaria* reveal the extent to which Coleridge relies on the body to execute the method that he outlines in *The Friend*. He asserts that 'the relation' between God and his creation is 'common to all' and, once it is 'known ... each [thing] becomes discoverable, together with its peculiar relations to its concentrics in the common sphere'.[92] It not only produces all other relationships in the phenomenal world, but it also makes them comprehensible to the human subject. The relationships that proliferate illustrate the originary relationship between God and the world, and it is no coincidence that 'the RELATION of objects is the prime *material* of Method, and that the contemplation of relations is the indispensable condition of thinking methodically'.[93] Coleridge's method also enables human beings to intuit the relationships between nature and God as well as all the others that comprise the phenomenal world.

Embodied Cognition

Coleridge's discussions of life, cognition, and knowledge production are so abstract that it can be difficult to envision how they apply to actual living, embodied beings. Coleridge's elision of the body in 'Theory of Life' and *Biographia* has encouraged scholars to read his work as a rejection of the physical in favour of disembodied laws, minds, and vital sparks. I attribute his erasure of the body as a response to the distress caused by his poor health, his opium addiction, and the anguish they caused. Perry connects Coleridge's eventual 'rejection of "the Body as Body"' to 'the appalling feeling of descent into materiality which accompanied his despair'.[94] Looking at his later notebook entries on the writing of St Paul, Webster claims that 'there is no doubt that Coleridge's personal pain and physiological disintegration had a significant impact upon the way he thought and wrote about Body and Soul'.[95] Indeed, scholars who recognize the body's presence in his work, despite its conspicuous absence, avoid replicating the Cartesian error that Coleridge sought to resolve. Once we see the body's absent presence, it attunes us to the impacts that his specific embodiment had on his work, thinking, and career. For example, Alan Richardson surmises that Coleridge's opium use—a response to chronic illness—played a part in the long delay between writing

[92] Coleridge, 'Essay V', *Friend*, I, p. 461.
[93] Coleridge, 'Essay V', *Friend*, I, p. 458.
[94] Perry, *Coleridge and the Uses of Division*, p. 46.
[95] Webster, *Body and Soul in Coleridge's Notebooks*, p. xxix.

and publishing 'Kubla Khan' because the poet was anxious about opium's 'capacity to suspend or counteract, even to destroy the will'.[96] Similarly, Robert Mitchell notes that 'his opium use called into question his most fundamental beliefs about the nature of the human will'.[97] Opium addiction, as most medical professionals today recognize it, is a bodily not a spiritual ailment, but without this understanding Coleridge's unruly body was a stumbling block in his quest to formulate a neat metaphysics in which the will triumphs over, but remains connected to, corporeality.[98]

Stanback argues that 'Coleridge's pained body contributed dynamically to his modes of ... aesthetic creation, and also to his insight into the fraught relationships between the mind and the body, the flesh and the imagination, and man and God'.[99] The unruly body may have been a sticking point in Coleridge's thinking about organicism, but it also proves to be a crucial starting point. Since his theories would not make sense otherwise (or would fall into Cartesian dualism), Coleridge necessarily assumes embodiment in life and mind. The perceptible world is only one part of the system that enables the law of life to individuate. The relationship between embodied life and its milieu is as important as living beings in his triplex system. For Coleridge, as for other eighteenth- and nineteenth-century thinkers who theorized about cognition, this puzzle presents as the problem of interaction, which includes questions about sensation and how, via perception, it leads to abstract thought. How, if there are no ideal forms to grasp, does the embodied mind conceptualize widely applicable principles? To formulate a comprehensive theory of organicism, Coleridge needed to think through the relationships between percipient human beings, the physical world, and the ground of being.

The solutions advanced by thinkers such as Darwin, Priestley, Thomas Beddoes, Humphry Davy, and Wedgwood hinge on the mind's ability to first synthesize and then abstract commonalities from actual perceived objects. The triangle is their paradigmatic example. Dualists such as Brown, for example, claimed that the ideal triangle must exist as something like a Platonic form. In *Observations on the Zoonomia of Erasmus Darwin, M.D.* (1798), he writes that people can 'have a general idea of the nature of triangles' only if 'their common properties may be objects of thought, without reference to particular degrees'.[100]

96 Richardson, *British Romanticism*, p. 50.

97 Mitchell, *Experimental Life*, p. 78.

98 For an analysis of Coleridge's complaints about his diminished pleasure response in light of contemporary research into the physiological aspects of addiction, see Thomas H. Schmid, '"Diminished Impressibility": Addiction, Neuroadaptation, and Pleasure in Coleridge', in *Romanticism and Pleasure*, ed. by Thomas H. Schmid and Michelle Faubert (Palgrave Macmillan, 2010), pp. 165–186.

99 Stanback, *Wordsworth-Coleridge and the Aesthetics of Disability*, p. 223.

100 Thomas Brown, *Observations on the Zoonomia of Erasmus Darwin,*

Claiming that the right triangle is an abstract 'idea [that we must take] for granted as previously existing', Brown classifies all ideas as 'phenomena of mind' that exist irrespective of real, corresponding objects in the world.[101] In contrast, thinkers who advocated theories of embodied cognition argued that abstract ideas are an amalgamation of the objects and events perceived and experienced over the course of one's life. Darwin, for instance, argues that 'we can be sensible to a number of ideas at the same time' and can 'form compound ideas' from these 'simple objects'.[102] Davy claims that 'all Triangles that we have seen are representations on the retina ... of a peculiar figure included by three right lines' and that the triangles that we observe in the world, 'being frequently raised in succession of association[,] will at length be raised synchronically & hence will a peculiar complex idea of a triangle exist in the Mind'.[103] However, both the dualist and the materialist accounts locate knowledge outside the subject, either in an immaterial realm or in the physical world.

In Coleridge's view, knowledge gleaned from the external world had the ability to awaken thinkers to divine principles that can only be intuited from within. To avoid perpetuating Cartesian dualism, he posits that sensory perception, or 'sensibility', as Perry calls it, 'may form a complicating, enriching liaison with the kinds of understanding provided by the abstracting intelligence'.[104] Put more simply and less tentatively, in Coleridge's theories the sensible body's interaction with its physical surround provides a crucial link between concrete and abstract thought, even if, as he claims, 'truth ... must be found within us'.[105] Rather than envisioning divine truths in terms of materiality or immateriality, Coleridge saw them as principles that could only be realized in the mind through the body's attentive engagement with the physical world. His notebook entries from the period just prior to completing 'Theory of Life' and publishing the *Biographia* offer insight into his thinking about the body that provides an intermediary between his so-called materialist phase and the body's putative disappearance from these texts. Indeed,

M.D. (J. Johnson and J. Wright for Mondell & Son, 1798), p. 101, Eighteenth Century Collections Online <link.gale.com/apps/doc/CW0106965553/ECCO?u=asecs&sid=bookmark-ECCO&xid=70b022cf&pg=1> [accessed 6 November 2008]. A descendent of the Common Sense School, Brown became well known after he published his response to Darwin's *Zoonomia* (1794), which he challenged on the mostly basis of its rhetoric and style of argumentation.

[101] Brown, *Observations on the Zoonomia*, pp. 203, 149.

[102] Erasmus Darwin, *Zoonomia, or, the Laws of Organic Life*, Cambridge Library Collection, 2 vols (Cambridge University Press, 2009), I, p. 89.

[103] Humphry Davy, 'MS. Notebooks' (before 1800), The Royal Institution of Great Britain, loc. HD/13/e.

[104] Perry, *Coleridge and the Uses of Division*, p. 55.

[105] Coleridge, 'Essay IX', *Friend*, I, p. 495.

reading the notebook entries in conjunction with *Biographia* reveals the body's centrality to his theory of mind and imagination.

Despite Coleridge's interest in and knowledge about the life sciences, his published work suggests a thinker more interested in formulating abstract laws than discovering how they function physically and physiologically. His private writings, however, tell another story. Notebook entries from the first decade of the nineteenth century show that Coleridge considered the body as crucial to cognition. For example, in a note from 1803 that wonderingly exclaims 'when I think how perishable Things, how imperishable Thoughts seem to be!', Coleridge connects their persistence with the body.[106] He continues, asking 'For what is Forgetfulness?' and answers his rhetorical question by remarking that the body serves as the storehouse of past thought.[107] All one need do to recall a forgotten memory is 'Renew the state of affection or bodily Feeling, same or similar—sometimes dimly similar / and instantly the trains of forgotten Thought rise from their living catacombs'.[108] The living catacombs are an unambiguous vehicle that metaphorizes the body. Calling it a catacomb, which can denote either a labyrinthine underground passage or an underground chamber filled with graves, foreshadows his discomfort with the body but also reveals its inevitable role in cognition. In the first definition, catacombs evoke the mysterious unconscious that, as Richardson argues, Coleridge worried could supersede the will. The second definition associates body, or physicality, with death. Despite the foreboding metaphor, a similar idea makes its way into *Biographia*, which, regardless of the body's obfuscation, reveals an unmistakably reciprocal relationship between the mind and the body.

Here, the anecdote of the servant who, upon falling ill, suddenly and unaccountably speaks ancient languages leads Coleridge to conclude, as he does in the earlier notebook entry, 'that all thoughts are in themselves imperishable'.[109] Under the influence of a fever, the servant parrots phrases in Hebrew, Latin, and Greek that, as her physician eventually learns, she heard repeatedly while in the employ of a pastor who read aloud from books written in these languages. The frequent repetition of the stimuli over several years indelibly impressed her with the sound of the words, although not with the knowledge of their meaning. Attributing the 'true origins' of this phenomenon to 'the impressions made on her nervous system', Coleridge cites this as 'both

[106] Samuel Taylor Coleridge, *The Notebooks of Samuel Taylor Coleridge*, ed. by Kathleen Coburn, 5 vols (Routledge & Kegan Paul, 1957), I, sec. 1575.

[107] Coleridge, *Notebooks*, I, sec. 1575.

[108] Coleridge, *Notebooks*, I, sec. 1575.

[109] Samuel Taylor Coleridge, *Biographia Literaria, or, Biographical Sketches of My Literary Life and Opinions*, ed. by James Engell and W. Jackson Bate, Collected Works of Samuel Taylor Coleridge, 7, 2 vols (Princeton University Press, 1983), I, p. 114.

proof and instance that reliques of sensations may exist in an indefinite time in a latent state' in the body.[110] Immediately after making this claim, he shifts attention from '*the body terrestrial*' to '*the body celestial*', declaiming in elevated language that the celestial body unites humans with God ('the living chain of causes'), sanctions 'free-will', and establishes the 'absolute *self*'.[111] The shift in diction, from a tone appropriate for the case study to the high language of the sermon, functions as a red herring that directs readers away from the strong connection between cognition and embodiment implied in this anecdote. When read as a palimpsest of his notebook entry, it attests that the body cannot be erased.[112]

Working out his ideas about imagination and fancy in an April 1811 notebook entry, Coleridge assumes that the body not only stores the relics of sensation but that it is also crucial to higher thought. He asserts that the 'Soul differences itself from any other Soul ... by *form* or body only'.[113] In other words, souls individuate only in conjunction with a physical body. Although Coleridge tries to discount physicality by claiming that 'all form as body ... is dead', he nevertheless establishes a necessary connection between embodiment, perception, and cognition.[114] Abstract thinking first requires the ability to distinguish one thing from another before we can synthesize manifold sense data into unified concepts. Without the body, which houses the 'forma efformans' (creative faculty), 'there would be no *fixation*, no distinct perception or conception'.[115] The body connects human beings to their environment by enabling them not only to perceive it but also to distinguish themselves from it. It 'gives consciousness to the Subject' by enabling it to recognize itself as different from but still part of a larger system.[116] In this notebook entry,

[110] Coleridge, *Biographia*, I, p. 113.

[111] Coleridge, *Biographia*, I, p. 114.

[112] Ironically, Coleridge uses this case study to undercut Hartley's doctrine of association. Christensen appositely points out the contradiction between his claim about freewill and the associationist logic that underwrites it. In Christensen's view, it belies Coleridge's assertion that 'the will, and the reason are perhaps never wholly suspended'. Coleridge, *Biographia*, I, p. 112. He likens the servant's feverish ramblings to Coleridge's marginal method of composing the *Biographia*—that is, translating, paraphrasing, or directly transcribing other authors' words without attribution—which, like the delirious woman, repeats others' signifiers without stable referents. Christensen sees it as an illustration of the deconstructionist tenet that 'No one can fully possess anything—certainly not language, least of all himself'. Christensen, *Coleridge's Blessed Machine*, p. 113. Ashley Miller analyses this passage to argue that Coleridge's poetic theory is suspicious of the fragment. See Miller, '"Striking Passages"'.

[113] Samuel Taylor Coleridge, *The Notebooks of Samuel Taylor Coleridge*, ed. by Kathleen Coburn, 5 vols (Routledge & Kegan Paul, 1973), III, sec. 4066.

[114] Coleridge, *Notebooks*, III, sec. 4066.

[115] Coleridge, *Notebooks*, III, sec. 4066.

[116] Coleridge, *Notebooks*, III, sec. 4066.

Coleridge is working out what he means by the term fancy. Later, in *Biographia*, he demotes fancy to a 'mode of Memory' and assigns its perceptive powers to the primary imagination, the 'prime Agent of all human Perception'.[117] This basic ability to gather sensations and perceptions into cohesive images—whether called fancy or imagination—is the first step in creating unified concepts that form the basis of abstract knowledge. Without the physical body, Coleridge suggests in his notebook entry, humans would never 'arrive at the knowledge of the living Being' of which it is a 'Symbol & outward & visible Sign'.[118] Setting aside quarrels about the relationship between symbols and signs and their referents, it becomes clear that the relational body is fundamental to Coleridge's concept of the imagination.

Coleridge's claim that 'All knowledge rests on the coincidence of the object with the subject' becomes physiologically legible through the processes that he describes in 'Theory of Life'.[119] Here, the way the lifeforce instantiates and structures organisms, particularly human beings, has implications for how embodiment factors into Coleridge's theory of cognition. The lifeforce's manifestation in living bodies parallels the process by which the physical world comes into and maintains being. Just as time and space enter into relationships to form increasingly complex forces, the poles of life—unity and individuation—enter into relationship with each other to manifest as increasingly complex organisms. According to Coleridge, all organisms 'ascend from the laws of attraction and repulsion, as united in gravity, to magnetism, electricity, and constructive power, till we arrive at the point representative of a new and far higher intensity'.[120] By associating living beings with Newtonian physics, he suggests a connection between the physical world and the physical body. Coleridge uses a double metaphor to analogize embodied life: 'the progress of Nature is more truly represented by a ladder, than by the suspended chain', but it also 'expands in concentric circles'.[121] Both the ladder and the circles represent taxonomic schemes. The ladder corresponds to vertical levels of organization based on physiological complexity that is akin to (but not the same as) contemporary biology's division of organisms into four categories: viruses, single-celled prokaryotes, single-celled eukaryotes, and non-colonial multicellular eukaryotes.[122] The circles correspond to something like family,

117 Coleridge, *Biographia*, I, p. 305.

118 Coleridge, *Notebooks*, III, sec. 4066.

119 Coleridge, *Biographia*, I, p. 252.

120 Coleridge, 'Theory of Life', *Shorter Works*, I, p. 538.

121 Coleridge, 'Theory of Life', *Shorter Works*, I, p. 537. The 'suspended chain' alludes to the Aristotelian great chain of being that was adopted in the early modern period.

122 Noel A. Heim and others, 'Hierarchical Complexity and the Size Limits of Life', *Proceedings of the Royal Society B: Biological Sciences*, 284.1857 (2017), doi:10.1098/rspb.2017.1039. Plants, animals, and people belong to the last category.

genus, or species. While Coleridge organizes life according to physiological structure, putting organisms with more complex physiology higher on the ladder, he retains the relationality of organicism with the concentric circles. The more sophisticated the physiology, however, the more sophisticated the cognitive capacities.

For Coleridge, animal life cannot be separated from sentience. For organisms higher on the ladder than plants and mostly immobile creatures, such as molluscs, Coleridge's theory of mind implicitly entails embodied cognition. As they ascend and complexify, animals develop a greater awareness of and increasing capacity to understand and manipulate their environments. In neurophenomenological terms, a creature's structure informs what it can perceive and how much it comprehends the world that discloses itself to it. Cognitive capacity depends on the degree of organization, which is determined by the relationship between the two poles of life. Unsurprisingly, the more individuation dominates, the more physiologically and cognitively sophisticated the species will be. For example, in polyps and molluscs, 'so languid is the power of individuation' that they all look nearly the same.[123] Moving up the ladder just a few rungs, 'we find the individuals separate, a more determinate form, and in the higher species, the rudiments of nerves, as the first scarce distinguishable impress and exponent of sensibility'.[124] Even animals on the lower rungs of the ladder have simple nervous systems that enable them to engage productively with their surroundings.[125]

In keeping with late eighteenth-century medical concepts, Coleridge classifies cognitive capacity in terms of reproduction, irritability, and sensibility. As they ascend the ladder, lifeforms exhibit increasing irritability (muscular sensitivity) and sensibility (nervous system sensitivity), which dictate their responsiveness to the environment. 'In the insect', for instance, 'we find the distinct commencement of a separation between the exponents of sensibility and those of irritability; i.e. between the *nervous* and *muscular* system'.[126] Being relatively low in individuation, insects have 'no confluence of nerves into one reservoir' and thus lack 'the independent existence of sensibility as sensibility'; their 'whole muscularity ... is the organ of irritability; and the

[123] Coleridge, 'Theory of Life', *Shorter Works*, I, p. 540.

[124] Coleridge, 'Theory of Life', *Shorter Works*, I, p. 540.

[125] Clearly, the cognitive activity and awareness of which animals and lower order organisms are capable differs vastly from the cognitive activity, including abstract thinking and self-consciousness, which characterizes neurotypical human thought. For arguments that warn against eliding species boundaries despite the challenge such a position poses to cognitive chauvinism, see Sha and Faflak, 'Introduction', pp. 9–17; Sha, 'Blakean Experience'. Coleridge, while perhaps implicitly challenging cognitive chauvinism in 'Theory of Life', explicitly maintains a firm boundary between human beings and other animals.

[126] Coleridge, 'Theory of Life', *Shorter Works*, I, p. 543.

nerves themselves are probably feeders of the motory power'.[127] Although Coleridge believed that insects lacked brains, here he recognizes a modicum of sensibility that helps them navigate their world. Lower order organisms, to which Coleridge did not attribute souls and which he was invested in distinguishing from human beings, have cognitive abilities in proportion to their organizational structures. In this way, clearly linking cognition to embodiment, he 'demonstrates his capacity to recognise that Body and Soul, Life and Mind, are distinct though truly interdependent aspects of the human being'.[128] At the same time, however, he defines cognition proper as the ability to think abstractly and to decompound and synthesize percepts and ideas.

In 'Theory of Life', the ascension of life culminates in the human body with not just its conscious but its self-conscious mind, which Coleridge describes as 'the apex of the living pyramid'.[129] The human body is nature's apotheosis because it comprises the perfect ratio of individuation to unity. The relationship between its poles is balanced, thus the lifeforce manifests at the height of sophistication. In human physiology, equal amounts of inward and outward striving energy enable people to be correspondingly attuned to both the inner and outer worlds. As a species, humankind perfectly balances outwardly focused objectivity with internally focused subjectivity. The 'force of organic power has attained an inward and centripetal direction', enabling introspection and abstract thought, and, even though Coleridge avers that knowledge of God or the Law comes from within, the human body stands at the gateway of the inner and outer worlds, granting access to both through its 'doors of perception'.[130] In *Biographia Literaria*, Coleridge asserts that 'sensation itself is a vision nascent, not the cause of intelligence, but intelligence itself revealed'.[131] Thus, sensory perception is the starting point that enables people to apprehend the ground of being. Sensory experience is not an efficient cause, but it can orient people towards the final cause.

Even though Coleridge wanted to avoid merging humans uncritically into the natural order, he did not want to divorce them as entirely as Descartes had. In a fragment titled 'Solitary) (Gregarious', he extends his thinking from 'Theory of Life', focusing exclusively on humankind. In it, he explores 'the connections which yet remain between the highest life form and those below'.[132] According to J.H. Haeger, Coleridge did not separate human beings

[127] Coleridge, 'Theory of Life', *Shorter Works*, I, p. 543.

[128] Webster, *Body and Soul in Coleridge's Notebooks*, p. 61.

[129] Coleridge, 'Theory of Life', *Shorter Works*, I, p. 550.

[130] Coleridge, 'Theory of Life', *Shorter Works*, I, pp. 550–551; William Blake, *The Marriage of Heaven and Hell* (Oxford University Press, 1975), fig. Plate 14.

[131] Coleridge, *Biographia*, I, p. 286.

[132] J.H. Haeger, 'Coleridge's Speculations on Race', *Studies in Romanticism*, 13.4 (1974), pp. 333–357 (p. 334).

entirely from animals and lower order organisms. Rather, he thought that humankind 'incorporates the attributes of lower beings as [it] rises above them'.[133] For example, he saw human sociality as a vestige of 'the physical unity of "cluster" animals in different form', and the 'separate but independent subsistence of "shoal animals" is reflected in the "corpus politicum" of the human community'.[134] Coleridge analogizes, not unlike Darwin in *Zoonomia* (1794), comparing human beings to hive insects and other animals that live in cooperative communities. What separates people from other species is the capacity for reason, which enables them to intuit the principles of organicism. This 'true and living knowledge proceeds from within' and can be known only through introspection.[135] Yet Coleridge makes it clear that the body and the world are required to help uncover this inner knowledge.

Cognition, in Coleridge's theory, functions as an organic subsystem. The external world and the body work in tandem to produce knowledge, not just about the physical world but also about the *tertium aliud*, the new thing that exceeds parts and wholes. In Coleridge's case, the third thing is knowledge of the divine. His account of the physical world in 'Theory of Life' reframes materiality such that it 'is not a threat to meaning, but its precondition'.[136] He imagines the physical world as a place of possibility, and he also reframes the body so that it is not an obstacle, despite his misgivings about his own body, but a means of coming to know God. Together, the body and its milieu create the conditions for creativity, reason, and knowledge of the divine. The problem, he states in an early notebook, is not the body or the physical world, but those who 'treat ... living being[s] as inanimate'.[137] He targets men of science who refuse to consider metaphysical and theological questions in their research, such as Lawrence who maintained that 'An immaterial and spiritual being could never have been discovered amid the blood and filth of a dissecting room'.[138] The world, in Coleridge's view, must be studied and understood for the access it provides to the divine law that governs it, that is, organicism. Being structured by organicism means that the body and the world exist in an interdependent relationship in which both are equally necessary to the other. Without a physical environment the body would not exist, and without human beings 'what', as Percy Shelley asks, 'were thou, and earth, and stars, and sea,

[133] Haeger, 'Coleridge's Speculations on Race', p. 335.

[134] Haeger, 'Coleridge's Speculations on Race', p. 335.

[135] Coleridge, 'Theory of Life', *Shorter Works*, I, p. 500.

[136] Donald R. Wehrs, 'Epilogue: Literary Theory and Cognitive Studies', in *Cognition, Literature, and History*, ed. by Mark J. Bruhn and Donald R. Wehrs (Routledge, 2014), pp. 243–252 (p. 248).

[137] Coleridge, *Notebooks*, III, sec. 4066.

[138] William Lawrence, *Lectures on Physiology, Zoology, and the Natural History of Man: Delivered at the Royal College of Surgeons* (Benbow, 1822), p. 7.

/ If to the human mind's imaginings / Silence and solitude were vacancy?'[139] Human beings infuse their milieu with significance because it provides the means to make meaning.

Embodied Method

While the 'Essays on the Principles of Method' in *The Friend* outline the process for gaining access to knowledge of divine principles, the concept of Romantic structures of engagement helps elucidate the process by which the interaction between body and world produces this knowledge. According to Coleridge, 'in every act of conscious perception, we at once identify our being with that of the world without us, and yet place ourselves in contra-distinction to that world'.[140] Thus, self-consciousness is the human capacity for dual awareness—to know that we are individuals who are embedded in an environment and also that we phenomenologically exist as discrete, autonomous beings separate from the surround on which we depend. Despite the basic awareness of self and other, to arrive at the truth of humankind's existential condition, at least according to Coleridge, takes much thought and contemplation. Furthermore, the search for truth must be carefully directed, hence his method. Meditation on one's relationship to the world, self-reflection, and observation of the mind's workings are how one becomes aware of this knowledge, which exists innately. Aware of interiority as well as exteriority, human beings are uniquely positioned to recognize themselves as part of a larger system that includes themselves, the natural world, and God. Romantic structures of engagement in the body and in the external world work together to produce this specific kind of knowledge.

Just as a person's cognitive structures interact with environmental structures to produce quotidian knowledge of the world, Romantic structures of engagement produce, in Coleridge's paradigm of cognition, inner knowledge of the divine. Coleridge calls these internal structures reason and understanding. Nature, both beautiful and sublime, and aesthetic objects, such as poetry, comprise the external structures. If, as I have argued, Coleridge approaches the instantiation of ideas in structural terms, the autopoietic concept of structural coupling offers a way to understand how this knowledge is produced. In enactive accounts of cognition, structural coupling describes the interactive relationship between an organism and its environment based on embodied

139 Percy Bysshe Shelley, 'Mont Blanc', in Mary Shelley and Percy Bysshe Shelley, *History of a Six Weeks' Tour Through a Part of France, Switzerland, Germany and Holland: With Letters Descriptive of a Sail Round the Lake of Geneva, and of the Glaciers of Chamouni* (T. Hookham, and C. and J. Ollier, 1817), pp. 171–183 (ll. 142–44).

140 Coleridge, 'Essay X', *Friend*, I, p. 497.

and external affordances. An organism moves through and interacts with its environment, both muscularly and perceptually, in ways that are peculiar to its physiology. Knowledge 'of the perceived world', according to neurophenomenologists, 'is constituted through complex and delicate patterns of sensorimotor activity'.[141] The organism-environment relationship is, however, mutual in that both contribute to meaning-making. As with all dynamic systems, an organism occupies a fixed state space that delimits its possibilities. The world offers a set of constraints and affordances, such as light waves or gravity, that demarcate the bounds of knowledge. Colour vision is the classic example. Animals with trichromatic vision, such as human beings, see colour differently from those with more or fewer colour processing channels in their eyes. In addition to eye structure, light waves also determine the colour combinations available to any percipient being. In this way, both an animal's body and the world around it produce its species-specific knowledge about itself and its environment.

Thus, the structural dispositions of the human body, in conjunction with its physical surroundings, produce knowledge about its inner and outer worlds. In phenomenological terms, these two structural loci mutually determine how the world discloses itself to human beings (or any species). To understand truth, one must recognize the discrete components that make up multeity, as well as the relationships that bind them together into a unified system. However, this skill must be cultivated; otherwise, people risk falling prey to the mindless routines of daily life that blunt them to apprehension of the divine. As the subset of the cognitive apparatus responsible for aesthetic experience, Romantic structures of engagement produce this intelligence. According to Coleridge, in addition to empirical knowledge about the world and its operations, humankind also has inner knowledge that confirms an intuitive sense that the world is not an end in itself but takes its origin in something else. In *Biographia*, he indirectly explains his method: 'to obtain adequate notions of any truth, we must intellectually separate its distinguishable parts .... But having done so, we must then restore them in our conceptions to unity, in which they actually co-exist'.[142] When inner and outer Romantic structures of engagement engage, the resulting experience can awaken knowledge of God, but these experiences must be sought, and contemplation must be cultivated. If these are undertaken mindfully, they can reveal knowledge of the 'Law'.

Coleridge's method explains how human beings cultivate this awareness by orienting themselves properly to both the external and internal world. Proper orientation entails a deep awareness of organicism, which is predicated on the principle of relationality. Coleridge's method offers a way to develop the capacity to deeply understand that God enables human existence. This

[141] Varela, Thompson, and Rosch, *The Embodied Mind*, p. 164.

[142] Coleridge, *Biographia*, II, p. 11.

consciousness is gained by paying sustained attention to the relationships among the people and things that comprise the physical world. Coleridge asserts that 'understanding … reflects the [external] objects *subjectively*, that is, [it] substitutes for inherent laws and properties of the object the relations the objects bear to its own particular constitution'.[143] While not as exalted as reason, the understanding recognizes relationality but only as it pertains to the subjective self. This capacity to perceive the principle of relationality on a small scale can lead the subject to a reasoned comprehension of the organicism that structures the physical world and bring about knowledge of the divine.

The 'RELATIONS of objects', says Coleridge, 'are the prime *materials* of Method'.[144] By objects, he means things, both inanimate and animate, in the physical world that can be perceived by the body. He divides them into 'two kinds of relation', theory and law.[145] Theory encompasses the relationships in the physical world that are available to the senses, while law encompasses the living truth that can be obtained by reasoning about the relationships that one observes. Crucially, these two poles—theory and law, external and internal, the physical and the divine—are legible only in relationship to each other. Sha's discussion of the dialectic between innocence and experience in William Blake's definition of consciousness provides a clarifying analogy. He claims that for Blake 'the meaning of consciousness must derive from the dialectical energy of their opposition.… Only after experience can we perceive fully the value of innocence'.[146] Likewise, interiority can be experienced only in contrast to exteriority, while the divine takes on significance only in relation to the mundane.

Attending to relationships is the path to knowledge, and, at the same time, it is the very knowledge that must be discovered. By contemplating relationships in the external world, 'as well as [by paying] careful attention to the conduct of the mind in the exercise of Method itself', a person eventually arrives at the knowledge that relationality, and not hierarchy, structures organicism.[147] Returning to the taxonomic metaphors of the ladder and concentric circles, the ladder accommodates hierarchy, as Armstrong claims, but only to a degree. The circles that ripple out from the centre better characterize the relational structure that originates with and emanates from God. According to Webster, Coleridge's later work, which draws on the logic of a 'Triune God', 'had effectively proved that no *fundamental* hierarchy existed between the ... sources of "Body" and "Soul", because they found their common being in the Divine'.[148] Relationality *is* the divine principle that he urges people to seek. For Coleridge,

[143] Coleridge, 'Essay IX', *Friend*, I, p. 491.
[144] Coleridge, 'Essay V', *Friend*, I, p. 458.
[145] Coleridge, 'Essay V', *Friend*, I, p. 458.
[146] Sha, 'Blakean Experience', p. 105.
[147] Coleridge, 'Essay VI', *Friend*, I, p. 466.
[148] Webster, *Body and Soul in Coleridge's Notebooks*, p. 55.

method is a way of orienting oneself to one's world, to oneself, and to one's God.[149] However, it must be practised until seeing relationality, especially relationship to the divine, becomes habitual, and this practice requires training both body and mind.

Coleridge's treatment of Jakob Boehme and Martin Luther in *Biographia*, *The Friend*, and elsewhere reveals the importance of mind–body calibration. Functioning as case studies, these discussions represent Boehme and Luther as untutored religious men, who sometimes 'mistake the tumultuous sensations of [their] nerves, and the co-existing spectres of [their] fancy, as parts or symbols of the truths which were opening on [them]'.[150] In his essay on Luther in *The Friend*, Coleridge notes that he 'suffered many and most distressing effects of indigestion and a deranged state of the digestive organs', which, along with his mental disposition, is 'abundantly sufficient to explain all his apparitions and all his nightly combats with evil spirits'.[151] In his marginalia in Boehme's *Works* (1764–81), he remarks that the mystic 'frequently ... mistake[s] the dreams of his own overexcited Nerves, the phantoms and witcheries from the cauldron of his own seething Fancy, for parts or symbols of a universal Process'.[152] The problem is lack of an education that teaches them how to consult their bodies to verify the insights generated by their minds. As Peter Cheyne notes, Coleridge 'later relates in *Table Talk* [when] the empirical checks of the sense of the intellectual ordinance of reason is absent, the whirls of associations become delirious or manic respectively'.[153] Coleridge's oblique allusions to the body make it clear that it works with the mind not only to generate divine insight but also to ensure the veracity of such knowledge.

Similar to Wordsworth's claims about the poet in the Preface to *Lyrical Ballads* (1800) and *The Prelude* (1850), Coleridge holds that some seekers of truth have 'deeper feeling, and a stronger imagination, than belong to most

[149] The idea that Coleridge's writing can help orient readers takes inspiration from Mitchell, who claims that Coleridge's life-manuals 'provide readers with a set of protocols that would allow for life experiences' that were 'both an experience of the disorienting nature of life, and (as a consequence) an experience that altered one's life'. Mitchell, *Experimental Life*, p. 93. Sha also provides inspiration with his contention that in *Biographia* 'Coleridge collapses imagination and method' and 'demands readers to think about their own methods as they think about the imagination'. 'Such thinking', he continues, 'was requisite to the development of mind'. Richard C. Sha, 'Romantic Physiology and the Work of Romantic Imagination: Hypothesis and Speculation in Science and Coleridge', *European Romantic Review*, 24.4 (2013), pp. 403–419 (p. 412).

[150] Coleridge, *Biographia*, I, p. 151.

[151] Coleridge, 'Essay II', *Friend*, I, pp. 138–139.

[152] Samuel Taylor Coleridge, *Marginalia*, ed. by George Whalley and H. J. Jackson, The Collected Works of Samuel Taylor Coleridge, 12 (Princeton University Press, 1980), VIII, pp. 28–29.

[153] Cheyne, '"Wild Activity of Thoughts"', p. 25.

of those, to whom reasoning and fluent expression might have as a trade been learnt in boyhood'.[154] Since feeling and imagination are innate capacities, some people have greater natural organic sensibility even than those who were trained as children to develop their minds and to attend to their bodies. Take Wedgwood, for example, who suffered from intense bodily ailments that significantly affected his mental states. Educated in part by Darwin, he exhibits hyperawareness of the relationship between the state of his body and the state of his mind. His journals attest to the close attention that he paid to each and his skill at distinguishing illusion from real percepts. Likening thinking and reasoning to a trade that can be practised and mastered, Coleridge suggests not only that they can be cultivated but that they must be properly aligned to accurately interpret insights gained from the embodied imagination. His remarks on cognitive development in his 'Essays on the Principles of Method' underscore this point. Methodical thinking begins 'with our cognitions as with our children'; first there is a 'period of aimless activity and unregulated accumulation' that 'in the mind of a rightly-educated person' is followed by a 'period of orderliness, of circumspection, of discipline'.[155] Thinkers such as Boehme and Luther, it seems, were unable to bring enough discipline to their thinking to distinguish between true revelation and bodily induced hallucinations. Aligning circumspection and discipline with training, Coleridge makes it clear that education, whether undertaken in childhood or adulthood, is necessary to the method and to the acquisition of truth.

An education in methodical thinking not only teaches a thinker to distinguish between inner vision and bodily disorders, but it does so by training the pupil to engage with the external world and to study the relationships among the multitude of phenomena that comprise it.[156] Coleridge explains:

[154] Coleridge, *Biographia*, I, p. 150.

[155] Coleridge, 'Essay X', *Friend*, I, p. 499.

[156] In the 1820s and 30s, Coleridge mentored a small group of young men, convening weekly discussions to instruct them in the method. John Sterling, one of the attendees of Highgate Thursdays, gave rise to some controversy in 1850s, thanks to competing biographies of him by J.C. Hare, the Cambridge tutor who introduced him to Coleridge, and Thomas Carlyle, who despised the poet and poetry in general. Both men, for their own personal reasons, portrayed the mild-mannered Sterling as weak willed and easily led astray. While Hare bemoaned Sterling's decision to leave his curacy, Carlyle regretted the influence of Coleridge's 'moaning singsong of … theosophico-metaphysical monotony', which he believed contributed to Sterling's death. Thomas Carlyle, *The Life of John Sterling* (Chapman and Hall, 1871), p. 50. In fact, Sterling gave up his curacy because he suffered from consumption, and he died for the same reason. However, in the wake of Hare's and Carlyle's biographies, he became the icon of a 'young man lost in the chaos of the nineteenth century' and its unfixed ideas about religion, science, and biblical exegesis. William Blackburn, 'Carlyle and the Composition of "The Life of John Sterling"', *Studies in Philology*, 44.4 (1947), pp. 672–687 (p. 674). To read about this

> Method becomes habitual by contemplating relationships amongst objects, observer (self), and others: METHOD, therefore, becomes natural to the mind which has been accustomed to contemplate not *things* only, ... but likewise and chiefly the *relations* of things, either relations to each other, or to the observer, or to the state and apprehension of the hearers.[157]

Although internal knowledge of God 'can never be infused or impressed' *ab extra*, it 'may be trained, supported, fed, [and] excited' by the external world.[158] The physical world aids the process of knowledge discovery by feeding and exciting the embodied mind with evidence of the divine principle that, if properly contemplated, trains reason to recognize truth. 'Self-election' into inner knowledge 'is achieved by directing the organic processes of the mind and body to the object which they will ingest, thereby actively shaping the self'.[159] Coleridge's method provides instruction on how to best direct these processes. Attending to how the world discloses itself phenomenologically awakens and strengthens the intuition that it is a unified multeity held together by the relationships of the parts to each other and to the whole. Sustained contemplation of the relationships among the parts of the human and natural worlds, in Coleridge's view, leads to knowledge of their ultimate relationship with God.

fascinating controversy, see Anthony John Harding, 'Development and Symbol in the Thought of S. T. Coleridge, J.C. Hare, and John Sterling', *Studies in Romanticism*, 18.1 (1979), pp. 29–48, doi:10.2307/25600166; Blackburn, 'Carlyle and the Composition of "The Life of John Sterling"'; Anthony John Harding, 'Coleridge as Mentor and the Origins of Masculinist Modernity', *European Romantic Review*, 14.4 (2003), pp. 453–466; and C. Richard Sanders, 'The Background of Carlyle's Portrait in *The Life of John Sterling*', *Bulletin of the John Rylands Library*, 55.2 (1973), pp. 434–458.

157 Coleridge, 'Essay IV', *Friend*, I, p. 451.

158 Coleridge, 'Essay IV', *Friend*, I, p. 500.

159 Leadbetter, *Coleridge and the Daemonic Imagination*, p. 46.

Embodied Aesthetics

Along with 'a salubrious life spent in the company of nature', Coleridge saw poetry as a way to train the mind into active habits of thought that could enable the reader to recognize the organic principle of relationality.[160] As an activity that engages both body and mind, reading poetry elicits visceral responses that train readers' habits of thought, for better or worse. In *Table Talk* (1835), Coleridge likens Samuel Richardson's novels to being in 'a sick room heated by stoves', while Henry Fielding's work invokes 'an open lawn on a breezy day in May'.[161] In *Biographia Literaria*, his condemnation of novels of sensibility, German drama, and the gothic also connects the embodied reading experience with its potential pernicious effects on readers' minds. Novels such as Richardson's *Clarissa* (1748) sink the mind into confusion and inhibit its capacity for rational thought. Indeed, *Clarissa* overstimulates the body with its 'loaded sensibility', while 'the minute detail, the morbid consciousness of every thought and feeling in the whole flux and reflux of the mind' recorded in the eponymous heroine's letters infect the mind with 'self-involution and dreamlike continuity'.[162] The novel's detailed representation of the reciprocal effects of physical and mental distress sinks the mind into lethargy. Coleridge criticizes German melodrama and gothic fiction, descendants of Horace Walpole's *Castle of Otranto* (1764), along similar lines, that is, they fail to 'bring … the whole soul of man into activity'.[163] Products of these genres, he claims, take their origin in 'the mere cramps of weakness, and orgasms of a sickly imagination on the part of the author', and thus succeed by appealing to 'the lowest provocation of torpid feeling on that of the readers'.[164] The language that he uses to describe these works explicitly locates their origins and their effects in bodily sensation. Yet they fail to stimulate the 'whole soul', not merely because they evoke 'cramps', 'weakness', and 'orgasms' but also because they draw too narrowly on a single mental faculty—the 'sickly imagination'. Novels that rely too heavily on eliciting physical responses use literary devices that primarily apply outer pressures on readers. Poetry, on the other hand, exerts a proportionate combination of outer and inner pressures that act equally on

[160] Modiano, *Coleridge and the Concept of Nature*, p. 1.

[161] Samuel Taylor Coleridge, *Table Talk*, ed. by Carl Woodring, The Collected Works of Samuel Taylor Coleridge, 14, 2 vols (Princeton University Press, 1990), I, p. 496.

[162] Coleridge, *Biographia*, II, p. 211. As much as he despised Richardson, Coleridge also admired him, much to his own chagrin. See *Biographia*, II, n. 3, p. 211 and *The Notebooks of Samuel Taylor Coleridge*, ed. by Kathleen Coburn, 5 vols (Routledge & Kegan Paul, 1961), II, sec. 2471.

[163] Coleridge, *Biographia*, II, pp. 15–16.

[164] Coleridge, *Biographia*, II, p. 211.

the body and mind, thus making it particularly suited to 'teach the science of Method'.[165]

For Coleridge, literature that appeals primarily to sensation directs readers' attention outwards for the sake of effect, but poetry directs it outwards with the goal of turning it inwards towards the truth. Poetry exerts external pressure by engaging the senses through sound, rhythm, metre, and visceral imagery as well as by the marks on the page. However, because these effects can be subtle, readers need to attend to them carefully and thoughtfully, which consciously trains their minds in analytical thinking, while they also experience the principle of relationality at the unconscious, bodily level. Metre recruits the body by creating rhythms, some of which are found in nature—the heartbeat, for example—that regulate speed and tempo. Metre, according to Coleridge, 'tends to increase the vivacity and susceptibility both of the general feelings and of the attention'.[166] Even if readers are not fully aware that it is happening, their attention is being solicited, encouraging more concentrated application. In contrast to the mental torpor induced by gothic literature, poetry's influence on the body stimulates the mind to greater awareness. Imagery also appeals to the mind through the body by drawing on mnemonic recall that requires readers to, perhaps subconsciously, recollect the sensory aspects of the thing being described in order to envision it mentally. 'From the rapid flow, the quick change, and the playful nature of thoughts and images', poetry requires 'perpetual activity of attention … on the part of the reader'.[167] Poetry also encourages active engagement with the external world by 'awakening the mind's attention from the lethargy of custom, and directing it towards the loveliness and wonders of the world before us'.[168] Once free of custom, the mind begins to become more fully aware of God's creation and if practised according to Coleridge's method it will turn from the external towards the inner expression of God's truth. Contrary to Webster's claim that 'by the time of *Theory*'s composition [Coleridge had come to] the conclusion that the human Mind was "self-contained" and did not need external objects', here we see how the right external stimuli can help the mind turn inwards towards a God that is knowable in both the mind and the world.[169]

Poetry also enables experiential engagement with ideas at play in a poem, offering a model of organicism that can be accessed via an embodied aesthetic

[165] Coleridge, 'Essay IV', *Friend*, I, p. 451.
[166] Coleridge, *Biographia*, II, p. 66.
[167] Coleridge, *Biographia*, II, pp. 21–22. See also a notebook entry from 1811, discussed below, which not only notes active attention as an effect of poetry but also attributes it to 'the rapid flow, & quick change, <and the playful nature> of the Thoughts & Images' it presents to the reader's mind. Coleridge, *Notebooks*, I, sec. 4115.
[168] Coleridge, *Biographia*, II, p. 7.
[169] Webster, *Body and Soul in Coleridge's Notebooks*, p. 8.

experience rather than one that is initially rational or intellectual. For Coleridge, a key to methodical thinking is not only to observe connections among objects but also to 'enumerate and analyze these relations' and to learn 'the conditions under which alone they are discoverable'.[170] In a well-written poem, the various parts—metre, imagery, word choice, and so forth—work together to create an experience that invites attentive readers to carefully examine the relationships among literary devices to understand how the poem produces its effects and conveys its meaning. As a 'species of composition … [that] is discriminated by proposing to itself such delight from the *whole*, as is compatible with a distinct gratification from each component *part*', a poem trains the mind by offering readers a miniature organic whole to decompound, examine, and re-synthesize.[171] According to enactive philosopher Ralph D. Ellis, art's 'effect depends on our directly *experiencing* rather than simply *intellectually acknowledging* the felt senses involved'.[172] Thus, readers must first experience before they can analyse, and good poetry encourages readers to engage with it intellectually because of the aesthetic response it arouses.

Indeed, Coleridge claims that good poetry carries readers forward 'not merely or chiefly by the mechanical impulse of curiosity, or by a restless desire to arrive at the final solution; but by the pleasurable activity of the mind'.[173] The mental pleasure derived from a poem should inspire readers to explore the source of that pleasure by enumerating and analysing it. It should motivate them to take a metacognitive approach so as to understand the workings of the poem. Observing the operations of one's own mind to trace how it disassembles and reassembles parts and wholes is another crucial aspect of Coleridge's method. As thinkers become conversant in this method, they begin to understand that not only do the parts of the poem work together to create meaning, but the poem also works with the embodied mind and its understanding of the external world. Coleridge's embodied aesthetics thus enable readers to deepen their sense of organic relationality as the core principle of the divine.

'The Eolian Harp': An Exemplar

In this final section, I offer a reading of 'The Eolian Harp' that demonstrates Coleridge's aesthetic method and reinforces his theory of embodied organicism. In contrast to many early twentieth-century critics, who found the poem intellectually insupportable because it contains unreconcilable philosophical

[170] Coleridge, 'Essay IV', *Friend*, I, p. 451.
[171] Coleridge, *Biographia*, II, p. 13.
[172] Ellis, *Curious Emotions*, p. 179.
[173] Coleridge, *Biographia*, II, p. 13.

contradictions, I argue that it withholds resolution because, for Coleridge, poetry's goal is not to state truths but to lead readers to their own discovery of it. 'The Eolian Harp' draws readers into its concrete, sensual imagery while also requiring metacognitive engagement to experience and understand it more fully. The images in 'The Eolian Harp' are external Romantic structures of engagement that evoke a response by involving the cognitive structures that are most attuned to aesthetics. Its iambic pentameter blank verse captivates readers at the pre-cognitive bodily level, while the poem's conceptual content engages the rational faculties. As it explores the implications of eighteenth-century theories of embodied cognition, 'The Eolian Harp' demands readers to think about the issues at stake and to draw their own conclusions by refusing to privilege any one position, including the one Coleridge offers.

Coleridge started writing 'The Eolian Harp', his first conversation poem, in 1795 as an effusive outpouring of love for his bride-to-be, Sara Fricker. When he published it a year later under the title 'Effusion 35', he had revised it to reflect his growing interest in physiology and cognition. Many critics read the pantheistic implications of embodied cognition that Coleridge explores in the poem as an extended heretical questioning that challenges traditional (that is, Cartesian) understandings of the relationship between God, nature, and human beings. Although some twentieth-century scholars 'admire the poem's structure', they find it indecisive and unsatisfying.[174] After laying out the idea of a one life that undergirds all organic life, Coleridge ostensibly retreats, chastened, back into orthodox religion, a move that, according to one critic, undermines 'all of the metaphysics expressed in the poem'.[175] Some mid-twentieth-century scholars such as M.H. Abrams, Albert Gérard, and Ronald Wendling see the final version of the poem, published in *Sibylline Leaves* (1817) under the title 'The Eolian Harp', as emblematic of Coleridge's thinking as it evolved during the first few decades of his career. It expresses, according to Abrams, 'what is most distinctive in Coleridge's mature thought and imagination'.[176] Reifying dualism, even if unintentionally, Abrams argues that 'The Eolian Harp' exemplifies Coleridge's transcendentalism by offering a glimpse of the 'observing consciousness ... freed from time and space which ... rises to a climactic moment of vision beyond sense'.[177] Other critics engage with the religious sentiment expressed; some see it as an illustration of Coleridge's theory of imagination; and others claim that it manifests his scientific ideas.[178] For example, McGann sees it as an exploration of 'ideas

[174] Wendling, 'Coleridge and the Consistency of "The Eolian Harp"', p. 26.

[175] William H. Scheuerle, 'A Reexamination of Coleridge's "The Eolian Harp"', *SEL: Studies in English Literature, 1500–1900*, 15.4 (1975), pp. 591–599 (p. 591).

[176] Abrams, *The Correspondent Breeze*, p. 164.

[177] Abrams, *The Correspondent Breeze*, p. 159.

[178] These critics tend to focus on the 'one life' lines, which they find puzzling because they

associated with the sensationalist controversy initiated by Locke's *Essay*' in the seventeenth century.[179] He argues that the early versions of the poem elicit an embodied response in readers, but in its final form it recapitulates the hyper-rational, disembodied logic of enlightenment.[180] Rather than reading the poem as 'an index to Coleridge's evolving thought and imagination', I contend that it provides an occasion for experiencing and practising the method that Coleridge develops in *The Friend*.[181]

In this, I build on more recent criticism by scholars such as Shelley Trower

present a position that Coleridge ostensibly had abandoned by 1817. Paul Magnuson suggests that critics have paid too much attention to these lines, and 'if "The Eolian Harp" is the seminal poem' that they claim it is, 'its influence must come from the earlier texts, not the later ones'. Paul Magnuson, '"The Eolian Harp" in Context', *Studies in Romanticism*, 24.1 (1985), pp. 3–20 (p. 3). Most twentieth-century scholars follow John Shawcross's assertion that Coleridge 'placed the exclusive transcendent consciousness of God above all other forms of consciousness'. Qtd in Wendling, 'Coleridge and the Consistency of "The Eolian Harp"', n. 18, p. 37. Invested in affirming Coleridge's professed transcendentalism, these critics hold that 'Pantheism was impossible to Coleridge', and they try to reconcile the contradiction between Coleridge's early Unitarian beliefs and his later Trinitarianism. Wendling, 'Coleridge and the Consistency of "The Eolian Harp"', n. 18, p. 37. Gérard, for example, claims that the 'one life' lines indicate that the poet had found a way to reconcile his experiences in nature with 'orthodox Christian transcendentalism'. Albert Gérard, 'Counterfeiting Infinity: "The Eolian Harp" and the Growth of Coleridge's Mind', *Journal of English and Germanic Philology*, 60.3 (1961), pp. 411–422 (p. 421). Wendling, on the other hand, sees the poem as an exploration of the mind that concludes that neither reason nor imagination sufficiently apprehends the 'one life' unless the poet remains connected to God and the human community. See Scheuerle, 'A Reexamination of Coleridge's "The Eolian Harp"', p. 591, for a brief critical history of early twentieth-century scholarship on the poem. For discussions of theology, see Mark L. Barr, 'The Lyric Dispensation: Coleridge, Mosaic Law, and Equivocal Authority in "The Eolian Harp"', *Studies in Romanticism*, 44.3 (2005), pp. 293–316; Joseph McQueen, '"Old Faith Is Often Modern Heresy": Re-Enchanted Orthodoxy in Coleridge's "The Eolian Harp" and "The Rime of the Ancient Mariner"', *Christianity and Literature*, 64.1 (2014), pp. 21–42. For a reconsideration of the auditor, see Catherine E. Ross, 'Coleridge's "The Eolian Harp"', *The Explicator*, 91.1 (2002), pp. 17–20. For an ecocritical perspective, see Timothy Morton, 'Of Matter and Meter: Environmental Form in Coleridge's "Effusion 35" and "The Eolian Harp"', *Literature Compass*, 5.2 (2008), pp. 310–335.

179 Jerome McGann, *The Poetics of Sensibility: A Revolution in Literary Style* (Oxford University Press, 1996), p. 21.

180 McGann, *The Poetics of Sensibility*, p. 20. According to McGann, as Coleridge revised the poem from an effusion into 'a conversation poem, "The Eolian Harp" evolved a critique of its initial enactment: a critique of the ideas of sensibility and, more crucially, the poetic practice (the effusion) licensed by those ideas'. McGann, *The Poetics of Sensibility*, p. 20. Where an effusion is a form that celebrates sensation and emotion, the conversation poem, McGann claims, condemns these embodied responses because they are beyond the pale of human consciousness.

181 Abrams, *The Correspondent Breeze*, p. 160.

and Michael Raiger that examines the poem within the context of eighteenth-century intellectual history. I read 'The Eolian Harp' as an attempt to train readers in Coleridge's method by drawing attention to relationality.[182] As the poem progresses, Coleridge takes readers through various ideas about the relationship between mind and matter that were in circulation at the time. He represents their potentially problematical implications—passivity, pantheism, and permeable subject–object boundaries—but he does not answer the philosophical questions that these theories raise. Rather, the poem enacts his method aesthetically by inviting readers to prepare themselves to better intuit truth by engaging with the phenomenal world, thus cultivating their thinking. The poem sets up a problem for readers to consider in order to strengthen the habit of methodical thinking.

'The Eolian Harp' is not merely a poem about cognition; it is one that requires readers to do a certain kind of thinking.[183] In my reading, the poem's insights are neither free from time and space, as Abrams claims, nor are they meant to be. Coleridge's representation of the mind does not exist independently of material constraints but acts in relationship to the external world. Both structurally and semantically, Coleridge situates an embodied mind within a concrete environment that stimulates and supports it. He offers a theoretical alternative to physiological accounts of cognition that avoids falling back into dualism. Yet Coleridge does not insist that his is the definitive solution, leaving the question open for readers to work through methodically so that they might arrive at the truth on their own. Furthermore, where many critics see the turn to religion near the end of the poem as a retreat into orthodoxy after dangerous flights of speculation, I argue that 'The Eolian Harp' privileges the relationship between the mind, body, and environment—regardless of what that might portend ontologically. The poem closes not with

[182] Trower considers the poem in the context of eighteenth-century nerve theory, arguing that the Hartleyan ideas that it expresses challenge the 'canonical understanding that Romantic poetry sought … to transcend the material body and world to reach an ideal or spiritual realm'. Trower, 'Nerves, Vibration and the Aeolian Harp', para. 3. Raiger troubles understandings of Coleridge's organicism, arguing that 'The Eolian Harp' 'presents a philosophical account of mind which is both passive and vital, material and spiritual …, whereby both mind and nature are organically linked, part of the same system of material operations brought into action by God's agency'. Raiger, 'The Intellectual Breeze', p. 83. These readings draw attention to important historical contexts that were overlooked during the latter part of the twentieth century.

[183] To wit, Marjorie Levinson argues that a 'process resembling thinking does happen in the kind of poem that has been classified as lyric'. Marjorie Levinson, *Thinking Through Poetry: Field Reports on Romantic Lyric* (Oxford University Press, 2018),, p. 262. Her point lends credence to my claim that Coleridge's theory of embodied organicism applies to work by other poets.

disembodiment or capitulation to religion but by underscoring embodiment as necessary to human cognition and to discovering God's truth.

Despite the many interpretations that claim 'The Eolian Harp' ends with religion, it in fact begins and ends with the body. From the very first line, Coleridge draws attention to both the body and the environment in which he situates it, thereby emphasizing the importance of embeddedness. The first stanza locates the characters, and by extension readers, in a physical landscape that he describes in abundant sensory detail. The poem opens with Sara's 'soft cheek reclined / thus on [the speaker's] arm', a move that establishes both the poem's atmosphere and its theme.[184] The physicality between the lovers in these lines evokes an intimate, warm atmosphere grounded in a cosy sense of touch. The lines that follow draw on nearly all the remaining senses as the speaker and the auditor 'watch the clouds, that late were rich with light', smell the 'exquisite … scents / Snatched from yon bean-field!', and listen to the 'stilly murmur of the distant Sea'.[185] The first line positions readers in the couple's loving embrace, while the rest of the imagery ushers them into an almost tangible landscape. This embodied warmth permeates the poem even in its more speculative moments. The first stanza invites the reader to introject the connections between the speaker, Sara, and nature so that even when the narrative moves away from her, the intimacy between the speaker and readers remains intact. Additionally, the opening scene affords the sensory stimuli that leads the speaker to philosophize later in the poem.

The first line also makes it clear that 'The Eolian Harp' is as much about the mind as it is about the body's interaction with the landscape. Coleridge represents the mind not only as embodied but also as productively engaged with the natural world. The speaker addresses the auditor as 'My pensive Sara!', positioning her as a thinking subject in meditation on the sensual scene in which she is situated, while the parenthetical asides that disrupt the speaker's descriptions of the landscape further highlight cognition by inserting reflective commentary into the sensual descriptions.[186] In addition to envisioning 'white-flowered Jasmin, and the broad-leaved Myrtle', for example, readers are told that these flowers are 'Meet emblems … of Innocence and Love!', and a few lines later they learn that wisdom should be as 'Serenely brilliant' as 'the star of eve'.[187] Coleridge not only directs readers' interpretation of these natural objects, but he also presents his audience with a split perspective that enables them to simultaneously inhabit the landscape and the speaker's mind.[188]

[184] Coleridge, 'Eolian Harp', ll. 1–2.

[185] Coleridge, 'Eolian Harp', ll. 6, 9–10, 11.

[186] Coleridge, 'Eolian Harp', l. 1.

[187] Coleridge, 'Eolian Harp', ll. 4–5, 8, 7.

[188] Perry makes a similar argument, claiming that the 'Coleridgean parentheses contribute

He represents exteriority and interiority as mutual; the world and the mind inform each other equally. Before each parenthetical aside, Coleridge lavishly describes some natural object that inspires these reflections. At the same time, the running commentary imbues these objects with additional meaning. While nature provides the stimuli and the occasion for exegesis, the tacit body mediates sensation.

Coleridge implies, rather than explicitly describes, the speaker's and auditor's bodies. He describes the contact between the two bodies as 'most soothing sweet', evoking the phenomenology of embodiment rather than the body itself.[189] This invites readers to imagine soothing sensations in their own bodies as the sensory details become increasingly lush and even sexual. The sweet touch intensifies the dreamy intimacy of new love, which is heightened by a 'desultory breeze' that caresses the Eolian harp (and, vicariously, readers' bodies) into 'a soft floating witchery of sound' reminiscent of 'honey-dropping flowers'.[190] The language in the second stanza has moved from sensuality to voluptuousness, intensifying the lovers' sense of pleasure in the landscape and in each other. The breeze's caress, a metaphor for a lover's transgressive touch, sweeps across the harp and invites 'sweet upbraiding, as must needs / Tempt to repeat the wrong!'[191] This coquettish reproof invites the breeze to 'boldlier' sweep the strings, which produce 'long sequacious notes' and 'delicious surges'.[192] This sexual imagery, clearly the language of a love poem, makes it evident that the body is no mere abstraction.

It is notable that Coleridge incorporates the 'one life' lines into this lush stanza, representing it as a synesthetic mingling of sensory data: 'A light in sound, a sound-like power in light'.[193] As the source of 'Rhythm in all Thought', the 'one life within us and abroad' connects the body to the mind, and even though these celebratory lines approach pantheism, which Coleridge follows to its logical conclusion in the next stanza, they represent embodied cognition in terms both joyful and exalted.[194] The 'one life', interpreted in the context of Coleridge's method, can be understood as divine knowledge, which is internal ('within us') and external ('and abroad'). These lines also contain Coleridge's

to a total effect, the sense that another kind of perception is always possible'. Perry, *Coleridge and the Uses of Division*, p. 147.

[189] Coleridge, 'Eolian Harp', l. 2.

[190] Coleridge, 'Eolian Harp', ll. 14, 20, 23.

[191] Coleridge, 'Eolian Harp', ll. 16–17.

[192] Coleridge, 'Eolian Harp', l. 18. This metaphor replicates a problematical narrative about the seduction and sexual conquest of women. While it reflects the social context in which Coleridge was writing, it does not alter my reading of the poem. Thanks to Sara Lampert for pointing out the troubling metaphor in these lines.

[193] Coleridge, 'Eolian Harp', l. 28.

[194] Coleridge, 'Eolian Harp', ll. 29, 26.

theory of cognition, in which access to inner truth must first be stimulated by observing the interrelatedness of phenomena in the outer world. However, in the next stanza, Coleridge moves on to explore potential problems with the 'one life'.

The third stanza presents the most common eighteenth-century concerns about embodied cognition, beginning with mental passivity and the threat to human will. As the speaker lies stretched 'on the midway slope' 'behold[ing] / The sunbeams dance, like diamonds, on the main', the inferred warmth of the sun, soft grassy hill, and sparkling sunbeams lull the speaker into a flight of fancy even as they reinforce the connection between his body and his mind.[195] In thrall to sensation, the speaker loses control of his thoughts, which 'uncalled and undetained' flit idly across his 'indolent and passive brain, / As wild and various as the random gales / That swell and flutter on this subject lute'.[196] Mental acts arise spontaneously in response to external stimuli unbidden and unchecked by his conscious self. The description of the gales shifts. No longer 'delicious surges', the 'wild and various ... random gales' now signal the threat of passivity.[197] These lines warn readers of the danger of yielding completely to sensualism. Abandoning oneself to the external world without critical thought risks never discovering inner knowledge. At the same time, as an exemplar of Coleridge's embodied aesthetics, these lines offer active readers an antidote to mental passivity by rousing their minds into thought.

A notebook entry about Shakespeare's *Venus and Adonis* (1593), written in 1811, explains how poetry can awaken readers' minds into activity. Poems stimulate and require 'perpetual activity ... on the part of the Readers' Attention', and

> The reader's Mind & ~~heart~~ Fancy are forced into too much action to sympathize with the merely *Passive* of our Nature—As little can the mind thus roused & awakened doze ~~under the~~ and be brooded up by indistinct ~~Dreams & enfeebling Emotions~~ Passions, as the low lazy Mist can creep upon the ~~Lake~~ surface while a strong Gale is driving the Lake on in waves and billows before it—.[198]

In contrast to the portrait of an embodied mind at the mercy of external circumstance, Coleridge contends that poetry can 'rouse' the mind into action and clarity. Rather than deny the '*Passive*' aspect 'of our Nature', Coleridge

[195] Coleridge, 'Eolian Harp', ll. 34, 36–37.
[196] Coleridge, 'Eolian Harp', ll. 39, 41–43. Although 'The Eolian Harp' was written before Coleridge had outlined the distinction between imagination and fancy in the *Biographia*, the description in these lines conforms to his definition of fancy. See Scheuerle, 'A Reexamination of Coleridge's "The Eolian Harp"', p. 596.
[197] Coleridge, 'Eolian Harp', ll. 18, 42.
[198] Coleridge, *Notebooks*, III, sec. 4115.

theorizes an aesthetics that shakes the mind from lethargy.[199] In this notebook entry, Coleridge reverses the image of the gale. While it can be wild and random, as in 'The Eolian Harp', it can also blow away 'the low lazy Mist' and 'indistinct passions'. Unlike Richardson's novels of sensibility, good poetry functions as 'a strong Gale … driving the Lake on in waves' that rouse and awaken the mind from its 'doze'. Even as Coleridge warns readers that the mind can fall into passive indolence, he stimulates their minds into action with his verse.

The next stanza tackles the ontological monism inherent in physiological theories of cognition and its corollary, pantheism. Contrary to scholars who focus on the heretical questions raised in these lines, which they claim drive the speaker back into religious dogma, I argue that the lines enact a flight from embodiment that proves more terrifying than pantheism. While Coleridge found pantheism repugnant and profane, he did not necessarily find it frightening. In stanza four, the speaker and his implied body disappear, as does the Eolian harp and the landscape that serves as it setting. Abstract generalities take the place of concrete details as the speaker wonders about 'all of animated nature', which overwrites Clevedon as the specific locale observed by the characters in the early part of the poem.[200] The harp propagates into an unspecified number of 'organic harps diversely framed' and the 'gales / that swell and flutter on this subject lute' become a 'vast … intellectual breeze'.[201] The move from specific to abstract is highlighted by Coleridge's use of a demonstrative pronoun to indicate a particular object—'*this* subject lute!'—in the lines immediately preceding the shift, which makes the concept concrete.[202] Additionally, the fourth stanza is structurally distinct from the rest of the poem. It is the only one that lacks tangible imagery which corresponds to specific aspects of the external world. The mixed metaphors in these lines—the harps are 'organic', the breeze is 'intellectual'—underscore the confusion they depict. Visually, the stanza stands out as the one with the fewest number of lines and as the only one that poses a question.

Like his 'idle flitting phantasies', in the penultimate stanza the speaker seems to be suspended above the grassy slope in an anxiety-producing freefall of thought.[203] This disembodied intellectual flight leaves the speaker feeling 'Wildered and dark'.[204] In contrast to interpretations that see the turn to religion as a way to overcome terrifying bewilderment and assert that the

[199] See also Coleridge's description of the 'water-insect' that 'wins its way up against the stream, by alternate pulses of active and passive motion'. *Biographia*, I, p. 124.

[200] Coleridge, 'Eolian Harp', l. 44.

[201] Coleridge, 'Eolian Harp', ll. 45, 42–43, 47.

[202] Coleridge, 'Eolian Harp', l. 43.

[203] Coleridge, 'Eolian Harp', l. 40.

[204] Coleridge, 'Eolian Harp', l. 63. Wildered, according to the *OED*, is a variant of

'conclusion of the poem seemingly depicts Coleridge rejecting *in toto* all of the metaphysics expressed in the poem', I contend that it reinforces the importance of the body to his theory of organicism, emphasizes his theory of embodied cognition, and realizes his embodied aesthetics.[205] The final lines of 'The Eolian Harp' stage a rhetorical return to the body. According to Fairer, 'embodiment serves as a reassuring repose after the poetic and philosophical risks the poem has taken'.[206] Yet, where Fairer claims that this return 'realign[s] the mind in a separate dimension', I argue that the body provides stability, a haven to which the speaker can and does return after his foray into disembodied mental flight.[207] By the end of the poem, the mind is no longer separate from the body as it was during the speaker's anxious questioning. In the face of wildering darkness, the speaker turns back to the phenomenal world. Just as the poem opens with Sara and the speaker in intimate physical contact, it closes with 'Peace, and this Cot, and thee, heart-honour'd Maid!'[208] The emphasis on physicality in the first and last lines indicate the importance of the body to Coleridge, despite its elision in most of his published works. Furthermore, embodiment offers not just refuge from potentially lawless flights of fancy but also a way to know and experience the divine. The 'Faith' to which Coleridge returns is one 'that inly feels'.[209] The word 'inly' refers not to an abstract inner truth but to an embodied faith bolstered by knowledge of the external world.

Some critics read the poem as 'a series of meditations and fantasies that are qualified, if not rejected outright, by Sara's admonition'. [210] Indeed, calling Sara a 'Meek Daughter in the family of Christ' who 'holily dispraise[s] / These shapings of the unregenerate mind' positions her as the reproachful voice of

bewildered. 'Wildered, Adj.', in *Oxford English Dictionary Online*, n.d. <https://doi.org/10.1093/OED/2179359083> [accessed 13 December 2016].

[205] Scheuerle, 'A Reexamination of Coleridge's "The Eolian Harp"', p. 591.

[206] Fairer, *Organising Poetry*, p. 182.

[207] Fairer, *Organising Poetry*, p. 181.

[208] Coleridge, 'Eolian Harp', l. 64.

[209] Coleridge, 'Eolian Harp', l. 60.

[210] Magnuson, '"The Eolian Harp" in Context', p. 18. Although it is beyond the scope of this analysis, I cannot help but note the misogynistic view of Sara that critics have taken. Scheuerle speculates in an aside that 'Sarah [*sic*] probably surmised that he [Coleridge] was rejecting all thoughts that lay outside of her personal beliefs'. Scheuerle, 'A Reexamination of Coleridge's "The Eolian Harp", p. 597. Gérard calls the turn in which 'Samuel gives way to Sara—a sad anticlimax'. Gérard, 'Counterfeiting Infinity', p. 414. Ross summarizes this critical camp as 'those who see "pensive Sara" as a shortsighted and grimly dogmatic impediment to the narrator's poetic genius'. Ross, 'Coleridge's "The Eolian Harp"', p. 17. Wendling notes that the critical distaste for the poem's conclusion likely stems from the 'patronizing attitude toward her'. Wendling, 'Coleridge and the Consistency of "The Eolian Harp"', p. 27; see also n. 5, p. 27. See Scheuerle, 'A Reexamination of Coleridge's "The Eolian Harp"', n. 1, p. 491 for a list of critics who read the poem through this sexist lens.

religion.[211] Yet, from the opening of the poem, Sara has been represented not as a mouthpiece but as a thinking subject. Her censure is an act that calls attention to her active mind. The word 'shapings' suggests the mind's power to shape and direct thoughts, while the adjective 'unregenerate' implies an act of will. If there is any need to repent, it cannot occur passively but must come from the subject's free choice. Rather than indexing a 'growing rift in Coleridge's philosophical thinking at this time', the ending does not signal apostasy.[212] Neither does it endorse any of the positions that it outlines nor erase the difficulties inherent in theories of embodied cognition. Rather, the poem requires readers to engage with the thorny questions surrounding the relationships between the mind, nature, God. Readers are encouraged to think through the issues, but more importantly they are shown a way to find their own answers and to have their own experiences with 'The Incomprehensible!'[213] The intimate connections between speaker and auditor, reader and speaker, mind and body, and between the embodied mind and its environment that Coleridge sets up in the first few stanzas attune an attentive reader to the principle of relationality. By withholding resolution in 'The Eolian Harp', Coleridge requires readers to actively engage with the poem, to feel and to think.

In Coleridge's aesthetics, poetry itself provides evidence of an active mind. Poems are the 'shapings' that emerge from the poet's mind, much like the music that emerges from the interplay of wind and harp. 'The Eolian Harp' is emblematic of much Romantic poetry because it requires readers to attend to the reading experience, to think, and to foster habits of active engagement with poetry and with the external world. Aesthetic experience, for Coleridge, promotes a particular way of approaching the world that allows readers to develop methodical thinking that enables them to apprehend God's truth, which he believed indwells in each human being. Not only does poetry provide a way to discipline and train the embodied imagination, but it also enables the production of knowledge. As readers think about the poem and the issues it presents, new ideas, thoughts, and interpretations emerge. Readers create a 'a tertium aliud' that is separate from Coleridge and the poem, something that the interaction of the reader and the poem co-create.[214] This enactive view of poetry mirrors Coleridge's epistemology in which knowledge of self, other, and God is generated by the subject's interaction with the world and by contemplating the relationships between objects. Coleridge's method offers poetry as

[211] Coleridge, 'Eolian Harp', ll. 53–55.

[212] Raiger, 'The Intellectual Breeze', p. 78.

[213] Coleridge, 'Eolian Harp', l. 59.

[214] Coleridge, *Biographia*, I, p. 300.

an antidote to passivity because he requires the reader to engage and, in turn, train the body to engage with the world around it to discern meaning, similar to the way we discover meaning in a poem.

Chapter Four

Ethical Organicism

William Wordsworth

In the Preface to *Lyrical Ballads* (1800, 1802), William Wordsworth inveighs against the effects of modernity because, he contends, industrialization causes an 'increasing accumulation of men in cities, where the uniformity of their occupation produces a craving for extraordinary incident, which the rapid communications of intelligence hourly gratifies'.[1] He worries that living in overcrowded cities and working in monotonous factory jobs blunts the mind and creates a need for more lurid stimulation to rouse it from its torpor. Rather than mere hysteria, Wordsworth's objections reflect a concern that nineteenth-century conditions of modernity exacerbated the effect of a Cartesian worldview, which conceptually reinforces dissociation between self and other. The combined effects of Cartesianism and the changes in the material conditions of daily life depress rather than nourish 'organic sensibility', that 'Great birthright of our Being'.[2] Sensibility, a common term in the medical literature of the time, in this context refers to people's innate ability to experience and engage with their world. Repetitive, mechanized motion alienates people from their surroundings and from their bodies because it encourages disengagement and mindlessness as a means to cope with the dullness of their occupation. Uncoincidentally, this relationship of mind to body to world replicates, on a

[1] William Wordsworth, 'Preface', *Prose Works*, I, p. 129, ll. 166–69. Unless otherwise indicated, I quote from the final published version of the Preface that originally appeared in *Poetical Works* (1850) and that Owen and Smyser reprinted in *The Prose Works of William Wordsworth* (1974). This version substantively follows the 1802 revision, with only minor changes to the text.

[2] Wordsworth, 'Preface', *Prose Works*, I, p. 127, l. 132; Wordsworth, *The Prelude: 1799, 1805, 1850*, ed. by Jonathan Wordsworth, M.H. Abrams, and Stephen Gill (W.W. Norton, 1979), 2.239. Unless otherwise indicated, I quote from the 1805 version of *The Prelude*.

physical level, Cartesian ontology and epistemology, which separates the mind from the body and world. In contrast, in *The Prelude* (1805) and other poems Wordsworth represents the world as a web of relations in which human and nonhuman beings are situated in mutually co-dependent relationships. His organicism invites people to ground themselves in embodied experience, and his poetry offers them the opportunity to do so.

Like Samuel Taylor Coleridge, Wordsworth based his understanding of the world on his phenomenological experience of it. However, where Coleridge vexed himself about how to account for God in an anti-Cartesian theology, Wordsworth's writings suggest that he was concerned about the effects of deeply rooted belief systems on how people shape their collective milieu, which Maurice Merleau-Ponty defines as 'our human setting, our physical, ideological and moral situation'.[3] For example, when people 'are taught from infancy that we were born in a state of inferiority to our oppressors, that they were sent into the world to scourge and we to be scourged', they accept the inequitable conditions of an exploitive economic system because they see it as natural.[4] Similarly, the largely unconscious conviction that the soul is separate from the body and its environment causes problems from the local to the national because it leads to attitudes and actions predicated on disconnection that sediment into problematical socio-political structures. It causes people to value the hope of an abstract future state that transcends the material world over the richness of a life grounded in phenomenality. Treating Cartesianism as a metaphor that structures perception, shapes how the world discloses itself to people, and affects the way people experience and make sense of themselves in relationship to it, Wordsworth objects to it on phenomenological and ethical rather than ontological grounds. He challenges it as a way of seeing, understanding, and relating to self, nature, and society, which is not only disheartening but discourages ethical engagement. Many of Wordsworth's writings indicate that, like Merleau-Ponty and later neurophenomenologists, he understood cognition as situated. In *The Prelude*, in particular, he represents knowledge as enacted in the interaction between percipients and their surroundings. Thus, Wordsworth's account of mental development advances a theory of embodied cognition that underwrites his model of organicism.

Despite Mark J. Bruhn's warning in *Wordsworth Before Coleridge* (2018) not to 'emphasize and valorize Wordsworth's representation of the embodied and environmentally situated mind', in both *The Prelude* and *The Excursion* (1814) Wordsworth represents the mind as embodied and embedded as he

[3] Maurice Merleau-Ponty, *Phenomenology of Perception*, trans. by Colin Smith, Routledge Classics (Routledge, 2003), p. 157.

[4] Wordsworth, 'Bishop Llandaff', *Prose Works*, I, p. 36, ll. 190–191. Wordsworth made this remark in the unpublished 'Letter to the Bishop Llandaff' (c. 1793).

considers various ways of relating to nature and to society.[5] Wordsworth's representation of cognition in both poems puts the mind in a continuous and mutually productive feedback loop between the body and its physical surroundings. Exploring cognition in a range of environments from the pastoral to the urban, his poetry highlights the structural compatibility between bodies and nature, the fit of the human mind to the external world and vice versa. This mutually constitutive relationship exemplifies Wordsworth's organicism, which is no mere abstraction but manifests in the relationships among every part of the phenomenal world. It represents not only how he experienced nature but also how he thought self-other relationships should be structured. Without instantiation organicism is meaningless in Wordsworth's view. His poetry represents an external environment that works in tandem with human subjects to produce thoughts, feelings, and actions that affect how individuals interact with their surroundings, and to demonstrate that if people experience and understand it in this way they can facilitate the process to the benefit of themselves and others.

In contrast to New Historicist scholarship, I contend that Wordsworth does not represent a 'mind that keeps its own inviolable retirement from the object world' to avoid being affected by it but advocates active engagement with one's surround.[6] Neither *The Prelude* nor *The Excursion* suggests that people should be impervious to external conditions. Interaction is fundamental to Wordsworth's ethics, according to Adam Potkay, who argues that Wordsworth 'rooted [it] in encounters with other things or persons that "melt into our affections" and make us feel an interest in them, a part of them, and a responsibility, however undefined, for them'.[7] Understanding and experiencing the world in terms of relationality rather than disconnection encourages people to feel interconnected with and ethically responsible for each other. Regardless of whether it is ontologically 'true', the mutual co-dependence that he sees in nature can foster a worldview that facilitates ethical engagement.

With respect to ethics, Potkay also claims that the 'chastening and subduing, making us tender or humane, would seem to be precisely the ethical aim or "worthy *purpose*" of Wordsworth's most sublime and beautiful verse'.[8] In '*The Prelude* as a Philosophical Poem', Bruhn perspicaciously argues that the poem sets out Wordsworth's ideas discursively while it simultaneously 'orchestrates experiential effects that verify his … philosophy' and 'leads readers to

[5] Mark J. Bruhn, *Wordsworth Before Coleridge: The Growth of the Poet's Philosophical Mind, 1785–1797* (Routledge, 2018), p. 27.

[6] Marjorie Levinson, *Thinking Through Poetry: Field Reports on Romantic Lyric* (Oxford University Press, 2018), p. 40.

[7] Adam Potkay, *Wordsworth's Ethics* (Johns Hopkins University Press, 2012), p. 5.

[8] Potkay, *Wordsworth's Ethics*, p. 101.

experience' it for themselves.[9] Even beyond *The Prelude*, Wordsworth's poetry often enacts the concepts it represents, bringing forth meaning formally as well as semantically and creating opportunities for readers to become aware of their own situated meaning-making. Thus, the formal structures in *The Prelude* and *The Excursion* provide an occasion to understand and to experience relationality. In this way, his verse challenges tropological Cartesianism on many fronts and replaces it with figures of relationality in an attempt to get readers to feel an ethical responsibility towards all parts of the world, from 'the philosophic mind' to the 'meanest flower that blows'.[10]

As with the figures discussed in the previous chapters, Wordsworth employs concepts that fall under the rubric of twenty-first-century 4e theories of cognition. Some proponents of 4e, such as neurophenomenologist Evan Thompson, look to Kant as an eighteenth-century thinker who anticipated contemporary theories, but since we know, by his own admission, that Wordsworth did not read Kant, I trace his understanding of situated cognition and relationality to Erasmus Darwin's *Zoonomia* (1794), David Hartley's *Observations on Man* (1749), and Baron d'Holbach's *Système de la Nature ou Des Loix du Monde Physique et du Monde Moral* (1770).[11] Although he was not a materialist, I argue that Wordsworth adopted and adapted aspects of these theories to represent his own understanding of the world.

Tropological Organicism

Wordsworth's organicism expresses his phenomenological sense of the world as an interdependent system in which the parts mutually specify each other. Unlike Coleridge, who obsessively tried to formulate a ground of being that could accommodate his commitment to relationality while keeping God distinct from his creation, Wordsworth showed little interest in ontology. In contrast to recent claims by Bruhn and others, such as Graham Davidson, that Wordsworth was a post-Cartesian substance dualist, I examine the poet's writing in the context of embodied cognition, building on the work of scholars

[9] Mark J. Bruhn, '*The Prelude* as a Philosophical Poem', in *The Oxford Handbook of William Wordsworth*, ed. by Richard Gravil and Daniel Robinson, Oxford Handbooks (Oxford University Press, 2015), pp. 397–413 (p. 412).

[10] William Wordsworth, 'Ode', in *Poems, in Two Volumes, and Other Poems 1800–1807*, ed. by Jared Curtis, The Cornell Wordsworth (Cornell University Press, 1983), pp. 269–277 (ll. 189, 205).

[11] *Système* was translated into English as *System of Nature; or the Laws of the Moral and Physical World* in 1795 by William Hodgson. It proved so popular that another edition came out in 1797. Hereafter, I refer to it as *System*, although it is likely that Wordsworth read it in French.

such as Alan Richardson and David Miall.[12] Bruhn's admittedly conjectural argument traces Wordsworth's ontological influences to his study of geometry, Alexander Pope, Dugald Stewart, Ralph Cudworth, and Friedrich August Nitsch, whose work he encountered before he met Coleridge. Bruhn argues that the 'two … central tenets of his mature philosophical poetry' are 'the "one life" that "animates" the "great system of the world" and the ontological "fitness" it guarantees between mind and nature', which he claims that Wordsworth 'must have picked up', 'in all likelihood, from Ralph Cudworth's *True Intellectual System of the Universe* [1678]'.[13] Bruhn revives a century-old dispute in which this topic was hotly debated by Arthur Beatty, Melvin Rader, and their successors.[14] Contrary to later twentieth-century scholars such as H.W. Piper who

[12] See, for example, David S. Miall, 'Wordsworth's "First-Born Affinities": Intimations of Embodied Cognition', *Poetics Today*, 32.4 (2011), pp. 693–715; Alan Richardson, 'A Beating Mind: Wordsworth's Poetics and the "Science of Feelings"', in *British Romanticism and the Science of the Mind*, Cambridge Studies in Romanticism, 47 (Cambridge University Press, 2001), 66–92; John Savarese, 'Wordsworth's Scattered Minds', in *Romanticism's Other Minds: Poetry, Cognition, and the Science of Sociability* (Ohio State University Press, 2020), pp. 103–133; and Lisa Ann Robertson, '"Swallowed Up In Impression": Humphry Davy's Materialist Theory of Embodied Transcendence and William Wordsworth's "Tintern Abbey"', *European Romantic Review*, 26.5 (2015), pp. 591–614, doi:10.1080/10509585.2015.1070343.

[13] Bruhn, *Wordsworth Before Coleridge*, p. 130. Jonathan Wordsworth distinguishes a 'belief in animated matter' and a 'belief in a universe permeated by *the One Life*'. Jonathan Wordsworth, *The Music of Humanity: A Critical Study of Wordsworth's 'Ruined Cottage' Incorporating Texts from a Manuscript of 1799–1800* (Harper & Row Print, 1969), p. 185. Later, he describes it as an instinctual 'joy' in the 'presence of a shared life-force'. Jonathan Wordsworth, 'Wordsworth's "Borderers"', in *English Romantic Poets: Modern Essays in Criticism*, ed. by M.H. Abrams, 2nd edn (Oxford University Press, 1975), pp. 170–187 (p. 175). Paul Magnuson offers a more recent and capacious definition, claiming that for 'Wordsworth the "one life" signified variously a figure of nature's active energy, an image of human community, and a source of sublime emotion'. Paul Magnuson, 'The Lake School: Wordsworth and Coleridge', in *The Cambridge Companion to English Literature, 1740–1830*, ed. by Thomas Keymer and Jon Mee, Cambridge Companions to Literature (Cambridge University Press, 2004), pp. 227–243 (p. 132). My view aligns most closely with Magnuson's position.

[14] The debate about Wordsworth's ontological commitments originated in the early twentieth century with Arthur Beatty, *William Wordsworth: His Doctrine and Art in Their Historical Relations*, 2nd edn (University of Wisconsin, 1927) in which he contends, according to his rival Melvin Rader, that 'Hartley and English associationism [were] the main and almost exclusive factor that influenced the philosophy of the mature Wordsworth'. Melvin Rader, 'The Transcendentalism of William Wordsworth', *Modern Philology*, 26.2 (1928), pp. 169–190 (p. 169). Rader countered Beatty by arguing that the poet makes a 'sweeping denial of the senses and the reason'. Rader, p. 182. Later critics such as H.W. Piper, *The Active Universe: Pantheism and the Concept of Imagination in the English Romantic Poets* (University of London, Athlone, 1962), and Keith G. Thomas, *Wordsworth and Philosophy: Empiricism and Transcendentalism in the Poetry* (UMI Research Press, 1989),

attribute Wordsworth's belief that 'all organized forms of matter [are] in fact alive and capable of feeling' to revolutionary French materialist philosophers and British Unitarians, or contemporary scholars such as Marjorie Levinson who 'hear[s] the Spinozistic echo' in Wordsworth's poetry, Bruhn sees the 'one life' as purely Platonic.[15] I contribute to this debate by setting aside ontology to argue that Wordsworth was more interested in phenomenology. Indeed, his comments in his private correspondence indicate a lack of interest in metaphysics. In an 1840 letter to Henry Crabb Robinson, Wordsworth writes, 'I never read a word of German metaphysics, thank Heaven! though I doubt not they are good diet for some tastes I feel no disposition to meddle', which, even if he made it in jest, suggests that he did not share Coleridge's obsession with the ground of being.[16]

Wordsworth's comments to Isabella Fenwick about the 1804 'Ode' on immortality further suggest his interest in phenomenology. He reassures her that he views the 'ingredient of Platonic philosophy' in his ode, that is, the 'notion of [the] pre-existence' of souls prior to their instantiation in bodies, as a poetic device.[17] Always striving to capture the human experience, he explains that the 'pre-existent state has entered into the popular creeds of many nations; and, among all persons acquainted with classic literature', which gave it 'sufficient foundation in humanity' to authorize him 'to make for my purpose the best use of it I could as a Poet'.[18] In conjunction with this remark, the ode's content suggests that Wordworth sought to capture the all too human feeling of disconnection from one's surround.

Explaining the inspiration for the ode, Wordsworth recounts an oft cited childhood memory. He tells Fenwick that:

tended to align the poet with materialism—or empiricism, as it is often mislabelled. More recently, scholars such as Simon Jarvis, *Wordsworth's Philosophic Song*, Cambridge Studies in Romanticism, 67 (Cambridge University Press, 2007), and Bruhn argue that Wordsworth subscribed to some form of dualism, and Potkay argues that Wordsworth eventually adopted Immanuel Kant's dualist idea that the sublime is a form of 'supersensible reason that transcends the natural world'. Adam Potkay, 'Wordsworth's Ethical Thinking', in *The Oxford Handbook of William Wordsworth*, ed. by Richard Gravil and Daniel Robinson (Oxford University Press, 2015), pp. 679–692 (p. 688). Graham Davidson, *The Intelligible Ode: Intimations of Paradise* (The Lutterworth Press, 2023), also traces Wordsworth's thinking to geometry and the Cambridge Platonists, with no more evidence than Bruhn offers.

[15] Piper, *The Active Universe*, p. 71; Levinson, *Thinking Through Poetry*, p. 103.

[16] William Wordsworth, *The Correspondence of Henry Crabb Robinson with the Wordsworth Circle, 1808–1866*, ed. by Edith J. Morley, 2 vols (Clarendon Press, 1927), I, no. 145. Wordsworth made this remark in response to *Blackwood's Magazine*'s accusation that Coleridge plagiarized *Biographia Literaria* (1817), which irked him.

[17] William Wordsworth, *The Poetical Works of William Wordsworth*, ed. by Ernest de Selincourt, 5 vols (Clarendon Press, 1963), IV, p. 464.

[18] Wordsworth, *Poetical Works*, IV, p. 464.

> I was often unable to think of external things as having external existence, and I communed with all that I saw as something not apart from, but inherent in, my own immaterial nature. Many times while going to school have I grasped at a wall or tree to recall myself from this abyss of idealism to the reality. At that time I was afraid of such processes. In later periods of life I have deplored, as we all have occasion to do, a subjugation of an opposite character.[19]

Few critics these days interpret this passage as a 'constantly reiterated insistence on the validity of the "immaterial nature" of his being'.[20] Although he does not identify it as such, Markus Poetzsch reads this note phenomenologically arguing that this 'fascinating passage locates the child and perhaps childhood itself at a cognitive and experiential crossroads' that is experienced as 'disquieting liminality'.[21] Charles Rzepka also reads it as a 'liminal moment' in which Wordsworth entifies, or acknowledges the objective presence of, 'the world of physical, inorganic "things"'.[22] Touching the wall does not, according to Rzepka, deliver ontological certainty but 'ratifies not only the existence of the stones that are touched, but of the self as a physical thing that touches'.[23] In Anna Foca's interpretation, this passage records an encounter with the Kantian imagination which, she claims, terrified the boy. He learns the imagination's power 'to create an alternate reality which may successfully masquerade as *reality*' and 'threatens to overpower the individual consciousness and to mask delusions as reality'.[24] Rather than signalling his ontological position, this recollection recounts a moment of frightening disorientation.

Wordsworth's claim that he 'communed with all that I saw as something not apart from me, but inherent in my own immaterial nature' alludes to the sense of relationality that underwrites his organicism, and was a remark consonant

[19] Wordsworth, *Poetical Works*, IV, p. 463.

[20] Marion Montgomery, 'Wordsworth and the Ghost in the Mind: The Seductions of Metaphor', *The Georgia Review*, 23.2 (1969), pp. 169–182 (p. 176). Davidson seems to be a rare exception. In *The Intelligible Ode*, he contends that the Fenwick notes stand as evidence that 'the external world began to assert its separate reality in the mind of the growing boy' that later solidified into a belief in a transcendent power that exists independently of the natural world. Davidson, *The Intelligible Ode*, p. 98.

[21] Markus Poetzsch, 'Between the "Abyss of Idealism" and the Hard Wall of "Reality": Lessons of the Wordsworthian Child', *New Review of Children's Literature and Librarianship*, 20 (2014), pp. 15–25 (p. 22), doi:10.1080/13614541.2014.863642.

[22] Charles Rzepka, 'From Relics to Remains: Wordsworth's "The Thorn" and the Emergence of Secular History', *Romanticism on the Net*, no. 31 (2003), para. 6, doi:10.7202/008696ar.

[23] Rzepka, 'From Relics to Remains', para. 3.

[24] Anna Foca, '"Let Me Be the Calm You Seek": Imagination as (Safe)House in Wordsworth's "Tintern Abbey"', *Bucknell Review*, 42.2 (1998), pp. 31–42 (p. 36).

with his views in 1843 when Fenwick recorded the conversation.[25] He dispels the notion of pantheistic idealism raised by his description of his nature as immaterial by opposing it to the phenomenal world and, in the next sentence, materialism. If as Poetzsch, Rzepka, and Foca argue, this account represents a terrifying and liminal moment in the child's life, then he touches walls and trees to reorient himself towards external reality. Like 'Samuel Johnson's famous kicking of a rock to refute Berkeleyan idealism', Wordsworth reassures himself that he and the world exist.[26] Having gone to the other extreme as a young man, he tells us he experienced 'subjugation of an opposite character'. Materialism is, of course, the 'opposite' of the 'abyss of idealism' and neither satisfactorily accounts for Wordsworth's, nor most people's, phenomenological sense of the world. The problem with idealism and materialism, according to Merleau-Ponty is that the former 'mak[es] the exterior immanent to [the subject]', while the latter 'subject[s it] to causal action'.[27] Both ontologies 'falsify the motivational relations existing between the external and internal worlds and make this relationship unintelligible'.[28] Another way to think about this is that both systems render 'the motivating-motivated structure' incomprehensible because they either reduce the world to a mental projection or fail to account for the sense of constrained freedom that characterizes human existence.[29] Dualism, which purports to accommodate both, dissatisfied Wordsworth because of the sharp separation that it imposes on the subject and the object. Considering perception from a phenomenological perspective obviates 'the if-then of causality since, as Merleau-Ponty puts it, motives act through their meanings'.[30] The emphasis is on meaning and meaning-making, which accounts for the world, body, and mind. Contemporary phenomenology, like Wordsworth's poetry, situates humankind in a relational field without doing away with the boundaries that keep all parts of the world separate while still acknowledging their interconnectedness.

While Wordsworth's recollection, like all childhood remembrances, was too far removed from the event to provide reliable insight into his youth, his

[25] Wordsworth, *Poetical Works*, IV, p. 463.

[26] Rzepka, 'From Relics to Remains', para. 2.

[27] Merleau-Ponty, *Phenomenology of Perception*, p. 423.

[28] Merleau-Ponty, *Phenomenology of Perception*, p. 424.

[29] Philip J. Walsh, 'The Concept of Motivation in Merleau-Ponty: Husserlian Sources, Intentionality, and Institution', *Journal of the History of Philosophy*, 60.2 (2022), pp. 303–336 (p. 333). Hartley, Darwin, and Priestley argued for what Hartley calls practical freewill, which captures the idea of constrained choice. However, most objectors were too invested in arguing for unconstrained freedom to will to attend to this detail. In addition to Coleridge, *Biographia*, I, see William Cockin, *The Freedom of Human Action Explained and Vindicated: In Which the Opinions of Dr. Priestley on the Subject Are Particularly Considered* ([n. pub.], 1791), or Thomas Brown, *Observations on Zoonomia*, p. 101, for examples.

[30] Walsh, 'The Concept of Motivation in Merleau-Ponty', p. 310.

comments nonetheless illuminate his thinking at the time he made them. Whatever his motivation in 1804, in 1843 Wordsworth saw the 'Ode' as a representation of 'particular feelings or *experiences* of my own mind on which the structure of the poem partly rests'.[31] It is notable that he considers his mental experiences as constitutive of the poem's structure. From an enactive perspective, poems do not spring autochthonously from the poet's head (or imagination) but rather emerge from the interaction between the poet and the poet's milieu. Structurally, a poem is a formal abstraction that conveys meaning and, at the same time, a concrete set of 'spatial arrangements between and among the words in the imaginary box on the page'.[32] Poetry is, therefore, an excellent example of how knowledge is enacted in the interaction between the percipients and the external world, as Wordsworth demonstrates in *The Prelude* and *The Excursion*.

Wordsworth's famous definition of poetry, in the Preface to *Lyrical Ballads*, as the 'spontaneous overflow of powerful feelings' that 'takes its origin from emotion recollected in tranquillity' further suggests that he sought to represent phenomenological knowledge as he explored how experience shapes our understanding of and relationship to our surround.[33] His qualification that emotion must be reflected and meditated on before it can elicit good poetry indicates an awareness that poetry represents a carefully mediated representation of the internal and external world. The poetry in *Lyrical Ballads* reveals his interest in things not as they are but as they seem to human beings. The 'appropriate business of poetry', he says in 'Essay, Supplementary' to *Poems* (1815), 'is to treat of things not as they *are*, but as they *appear*; not as they exist in themselves, but as they *seem* to exist to the *senses*, and to the *passions*'.[34] Wordsworth distances himself from the first spontaneous flow of emotion so that, like a phenomenologist, he can examine the structure of experience, how emotion shapes meaning, and how meaning drives motive. His 'pursuit is not a passive registration of the impact of affections', asserts Nancy Yousef, 'but a formal practice involving attention'.[35] Poems such as 'Goody Blake and Harry Gill', 'The Thorn', and even 'Lines Written Above Tintern Abbey' represent reality as it is '*constituted* through the activity of the perceiving subject'.[36] As Wordsworth explored 'how mind and world are joined together' and how that

[31] Wordsworth, *Poetical Works*, IV, p. 463.

[32] Levinson, *Thinking Through Poetry*, p. 245.

[33] Wordsworth, 'Preface' to *Lyrical Ballads*, *Prose Works*, I, p. 149, ll. 627–629.

[34] Wordsworth, 'Essay, Supplementary to the Preface', *Prose Works*, III, p. 63, ll. 23–27.

[35] Nancy Yousef, 'After-Affects and Second Thoughts: Wordsworth, Eliot, and the Forms of Emotional Thinking', in *Romanticism and Consciousness, Revisited*, ed. by Richard C. Sha and Joel Faflak, Edinburgh Critical Studies in Romanticism (Edinburgh University Press, 2022), pp. 144–166 (p. 156).

[36] Walsh, 'The Concept of Motivation in Merleau-Ponty', p. 316.

affects perception and understanding, he put forward a theory of mind that underwrites his organicism.[37]

Situated Cognition

Where 'there is no indication that Wordsworth ever read' the Cambridge Platonists, scholars know that he was well versed in British and French materialist accounts of mind and nature.[38] His ideas about the mind reformulate theories of situated cognition that he encountered in work by Hartley, Joseph Priestley, Darwin, and Holbach. Although he did not keep meticulous notes, his letters, library holdings, and published works indicate that he read writings by these men. W.J.B. Owen and Jane Worthington Smyser note similarities to Hartley's and Priestley's thought in the Preface to *Lyrical Ballads*, particularly the definitions of 'philosophical language', while Duncan Wu thinks that he read *Observations* sometime between 1800 and 1808.[39] An explicit reference to Priestley in Wordsworth's unpublished 'Letter to the Bishop Llandaff' (c. 1793) suggests that he had read some of Priestley's work, either *Disquisitions Relating to Matter and Spirit* (1777) or his commentary on Hartley's theory of mind published in 1775.[40] While he was working on *Lyrical Ballads* in 1798, he asked Joseph Cottle 'to send Erasmus Darwin's two-volume, 1300-page medical treatise, *Zoonomia*' as quickly as possible.[41] According to Piper, Wordsworth's 'pantheistic creed involving a living, active universe' was heavily influenced

[37] Sam Hushagen, 'Sense Data and the Romantic Myth of the Given', *The Wordsworth Circle*, 53.3 (2022), pp. 311–133 (p. 313), doi:10.1086/720738.

[38] Davidson, *The Intelligible Ode*, p. 18.

[39] Wordsworth, 'Commentary: Preface', *Prose Works*, I, p. 169; Duncan Wu, *Wordsworth's Reading 1770–1799* (Cambridge University Press, 1993), p. 103. Wu draws on epistolary evidence from 1808. However, Hartley's work was commonly read at Oxford when Wordsworth was a student there. His library at Rydal Mount contained an English translation of Hermann Pistorius's German translation of *Observations* published in 1791. Chester L. Shaver and Alice C. Shaver, *Wordsworth's Library: A Catalogue* (Garland Publishing Print, 1979), p. 333. See also Will Christie, 'Wordsworth and the Language of Nature', *Wordsworth Circle*, 14.1 (1983), pp. 40–47; John Hayden, 'Wordsworth, Hartley, and the Revisionists', *Studies in Philology*, 81.1 (1984), pp. 94–118; and Jonathan Lamb, 'Hartley and Wordsworth: Philosophical Language and Figures of the Sublime', *MLN*, 97.5 (1982), pp. 1064–1085.

[40] Charles Pittman, 'An Introduction to a Study of Wordsworth's Reading in Science', *Furman's Studies*, 33.5 (1950), pp. 27–60 (p. 55).

[41] Richard Matlak, 'Wordsworth's Reading of *Zoonomia* in Early Spring', *Wordsworth Circle*, 12.2 (1990), pp. 76–81 (p. 76). See also Gavin Budge, 'Erasmus Darwin and the Poetics of William Wordsworth: "Excitement without the Application of Gross and Violent Stimulants"', *Journal for Eighteenth-Century Studies*, 30.2 (2007), pp. 279–308 (p. 279), doi:10.1111/j.1754-0208.2007.tb00337.x; Wu, *Wordsworth's Reading*, p. 72; Alan Bewell, *Wordsworth and the Enlightenment: Nature, Man, and Society in the Experimental Poetry*

by the French *philosophes*, including Holbach's *System*, which he encountered in France in the early 1790s.[42] Wu corroborates Piper's claim, venturing that Wordsworth read Holbach's four-volume work sometime in 1791 or 1792, and he had a 1791 French reprint of the book in his library at Rydal Mount.[43] As I will show, Wordsworth's representations of the mind, whether he was considering it in its developmental stages, as in *The Prelude* and *The Excursion*, or in the act of thinking, as in lyrics such as 'Lines Written Above Tintern Abbey', illuminate his ideas about organicism.[44]

Indeed, the challenge to Cartesian dualism advanced by Darwin, Priestley, and Holbach helped Wordsworth think about phenomenality. They thought about matter in a radically different way from the widely held eighteenth-century view of it as 'necessarily a *solid* or *impenetrable* substance, and naturally, or of itself, destitute of all *powers* whatever'.[45] Rather than envisioning matter as inert, a cornerstone of Cartesian ontology, they offer an account of matter as active, inseparable from the forces that enable it to structure itself. All three men invoke Occam's razor, arguing that substance dualism is unnecessarily complicated. Holbach asks, 'Would not it have been more natural to conclude, that since man, who is matter, and who has no ideas but of matter, enjoys the faculty of thought, that matter can think or is susceptible to the particular modification which we call thought?'[46] Philosophically and theologically, the Cartesian separation of the intellect from the human body was relatively new in the eighteenth century. Prior to the seventeenth century, people did not think that body and soul were separate but saw the latter as a lifeforce that structured the body. All living things had souls, so to speak. Davidson notes that 'classical and medieval thinking tended not to distinguish powers of the mind from their insights into the structures of the physical world'.[47] Yet in the

(Yale University Press, 1989); Richardson, 'A Beating Mind'; and Desmond King-Hele, *Erasmus Darwin and the Romantic Poets* (Macmillan Press, 1986).

42 Piper, *The Active Universe*, p. 68. Although Holbach was born in Germany, he was reared by his uncle in Paris. Michael LeBuffe, 'Paul-Henri Thiry (Baron) d'Holbach', in *The Stanford Encyclopedia of Philosophy*, ed. by Edward N. Zalta and Uri Nodelman, Spring 2020 (Metaphysics Research Lab, Stanford University), sec. 1 <https://plato.stanford.edu/archives/spr2020/entries/holbach/> [accessed 3 April 2021]. For an account of Wordsworth's time in France and his materialist influences, see Piper, *The Active Universe*, pp. 60–88.

43 Wu, *Wordsworth's Reading*, p. 74; Shaver and Shaver, *Wordsworth's Library*, p. 125.

44 Levinson defines lyric poetry as the enactment of 'thought happening or consciousness occurring'. Levinson, *Thinking Through Poetry*, p. 281.

45 Joseph Priestley, *Disquisitions Relating to Matter and Spirit* (Kessinger Publishing, 2005), pp. 2–3.

46 Holbach, *System*, I, p. 167.

47 Davidson, *The Intelligible Ode*, p. 15.

wake of John Locke's sensationalist psychology and Cartesian and Newtonian physics, the relationship of mind to matter had come under constant debate.[48]

Claiming that Cartesian theories of matter are based on 'no more than *superficial appearances*, and have therefore led to superficial and false judgments', Priestley offers a detailed physics of active matter based on Newton's concept of action-at-a-distance and drawing on the theories advanced by Roger Joseph Boscovich, who reached similar conclusions about matter.[49] Priestley advanced an early form of atomic theory which conjectures that solid objects are composed not of densely packed substantial matter but of small particles structured by the active forces inherent in them. His explanation of light in *The History and Present State of Discoveries Relating to Vision, Light, and Colours* (1772), which he quotes in *Disquisitions*, proposes that matter 'consists of physical points only, endued with the powers of attraction and repulsion, taking place at different distances'.[50] While Priestley was largely mechanistic in his approach, he provided a framework for the active view of nature associated with Romanticism. Darwin also advanced a theory of matter that more closely resembles the 'one life' espoused in Wordsworth's early poems. Approaching the Cartesian problem from the perspective of life sciences, he speculates, rather wildly, that sensorial power, a subtle energy that exists inside and outside of the body, enables organisms to initiate movement without the aid of an immaterial soul. In *Zoonomia*, he offers a truncated account of matter similar to Priestley's explanation, which provides the foundation of sensorial power's fantastic malleability.

Paul-Henri Thiry, Baron d'Holbach, an avowed materialist and atheist, explicitly challenged substance dualism in *System*.[51] Published under the name Jean-Baptiste de Mirabaud, *System* was popular among the revolutionaries who influenced Wordsworth's thinking, including the Girondists, John 'Walking' Stewart, and William Godwin.[52] He censures Descartes for being the 'first man who established that, "*that which thinks ought to be distinguished from*

[48] See John W. Yolton, *Thinking Matter: Materialism in Eighteenth-Century Britain* (University of Minnesota Press, 1983).

[49] Priestley, *Disquisitions*, p. 4.

[50] Joseph Priestley, *The History and Present State of Discoveries Relating to Vision, Light, and Colours* (J. Johnson, 1772), p. 391.

[51] From the 1750s to the 1780s, Holbach hosted salons for intellectuals who wanted to talk about ideas too radical to discuss in most public settings, including dinner parties. Despite the atheistic and revolutionary tendencies of his guests, Holbach 'managed ... to keep his coterie firmly in the mainstream of European society' and maintained his own place in polite society irrespective of his unconventional ideas. LeBuffe, 'Paul-Henri Thiry (Baron) d'Holbach', sec. 1.

[52] Richard Gravil, 'Tintern Abbey and *The System of Nature*', *Romanticism*, 6.1 (2000), pp. 35–54 (p. 41), doi:10.3366/rom.2000.6.1.35.

matter"'.[53] Rejecting the concept of an immaterial substance, Holbach asserts that the 'different properties' of matter, '—its different combinations—its various methods of action', enable it to take an 'infinity of forms [that] receive and communicate incessantly ... various motion'.[54] Motion, in his view, inheres in matter; everything in the phenomenal world is in constant motion. The forms that result from the various combinations of matter dictate organisms' and entities' function. Although, less detailed than Darwin, Holbach theorizes that the world and its species developed through evolution. Positing that kinetic matter existed in chaos, he states that 'the different matter of which the universe is composed must have, from all eternity, ... gravitated towards a centre, dashed against each other, met, been attracted and repelled, combined and separated'.[55] Physical organization, Holbach claims, developed 'as a necessary consequence of the disentangling of our globe'.[56] Again, akin to Darwin, he envisions matter organizing itself into the world that human beings now inhabit, continuing to organize itself in an eternal, self-perpetuating set of processes. Wordsworth encountered these theories as a young adult and, according to Piper, his concept of the 'one life', the idea that 'apparently inanimate objects are not "dead" but act by their own forces', originated with French and British philosophers among whom the belief was common.[57] These theories made it possible for Wordsworth to conceive of the physical world as active and to envision how mind and matter might be integrally connected, even if he did not agree or concern himself with their speculative science. Envisioning nature as an interdependent system, these thinkers contributed to Wordsworth's concept of it as an organic web of relations in which human beings are embedded.

Wordsworth's representation of the mind in many of his poems, but especially *The Prelude* and *The Excursion*, belies Bruhn's claim, and those advanced by earlier scholars, that he subscribed to a dualistic, 'ready-made framework of Christian association'.[58] Bruhn argues that Wordsworth does not see thoughts (mind) and feelings (body) as 'tautological but as systematic: the respective capacities are coordinated, even interpenetrating, but they are not coextensive'.[59] Yet Wordsworth consistently depicts cognition as a process that reciprocally implicates the body, the environment, and the mind. For him, the mind is embodied and embedded in its physical substrates as mental experiences are interwoven with the body as conscious and unconscious

[53] Holbach, *System*, I, p. 166. Holbach and Coleridge criticize Descartes in strikingly similar ways. See Chapter 3.
[54] Holbach, *System*, I, p. 31.
[55] Holbach, *System*, I, p. 57.
[56] Holbach, *System*, I, p. 147.
[57] Piper, *The Active Universe*, p. 69.
[58] Wordsworth, *The Music of Humanity*, p. 239.
[59] Bruhn, *Wordsworth Before Coleridge*, p. 29.

feeling-inflected memories. The structural and verbal patterns in *The Prelude*, for example, clearly stake a claim for situated cognition, beginning with the opening lines. In each major version of the poem (1799, 1805, and 1850), Wordsworth asks, 'Was it for this / That one, the fairest of all rivers, loved / To blend his murmurs with my Nurse's song[?]'[60] This question immediately introduces the twofold environment in which the human mind is embedded—nature and society. The following lines represent a mind mediated by the body and structurally interdependent with its surround. Derwent's 'ceaseless music through the night and day' and 'its steady cadence' are not a mere backdrop but partially constitute the child's mind.[61] The speaker states, in non-figurative language, that the river 'composed my thoughts', giving him 'a knowledge' not just of nature but of its interrelatedness with human beings.[62] The nurse's song highlights the web of human relationships, and the nurse's role in forming the child's mind underscores the importance of society. This passage, which opens the 1799 two-book version of *The Prelude*, also emphasizes the body with the introduction of the 'naked boy'.[63] The adjective 'naked' calls attention to the boy as an embodied being. Wordsworth's language in these opening lines is straightforward. It sets the scene of his early childhood and establishes the fundaments of his cognitive theory. In later versions of the poem, Wordsworth moves the apostrophe to the River Derwent a few hundred lines into the first book, but the key terminology remains largely the same.[64]

The rest of the verse paragraph bolsters the mind–body–environment relationship established in the first twenty or so lines. In addition to representing how the boy's physical environment moulded his mind, it replicates similar conditions for readers as they enact the knowledge in these lines with sensory images that engage their minds via their bodies. Juxtaposing the burbling river and the nurse's song with 'silent pools', Wordsworth amplifies the aural imagery.[65] The contrast calls attention to the language of sound, making it more noticeable and evocative. He also uses strong visual and tactile images to demonstrate the boy's visceral connection to the world around him, which populates readers' imaginations with warmth, colour, and texture. It is not difficult to picture him as he 'Basked in the sun, or plunged into thy streams' because Wordsworth contrasts the sun's warmth with the shock of

[60] Wordsworth, *Prelude*, 1.3 (1799).

[61] Wordsworth, *Prelude*, 1.9–10 (1799).

[62] Wordsworth, *Prelude*, 1.11, 14 (1799).

[63] Wordsworth, *Prelude*, 1.18 (1799).

[64] The 1850 version of *The Prelude* omits the 'naked boy'. This edition of the poem is curious because it replaces some visceral language with pious religious idiom but leaves much of it untouched.

[65] Wordsworth, *Prelude*, 1.18.

cold water.[66] According to Elaine Scarry, dyadic structures are particularly effective at engaging the imagination because they set the mind in motion. Another pair of images in these lines evokes visual motion. The juxtaposition of 'yellow grunsel' and the mountain in the distance draws the inner eye up from the ground progressively, first to the 'crag and hill', then to 'The woods' and 'Skiddaw's lofty height', and finally back to the middle distance, the field in which the 'naked savage' (the boy) stands in a 'thunder-shower'.[67] Scarry, who tracks a similar pattern in 'I Wandered Lonely as a Cloud' (1807), asserts that this poetic device creates 'motion [that] is felt somatically throughout the entire body'. Readers 'visualize, then incorporate, the moving picture so that there is a felt lift of the body'.[68] Opposing bright colours like the 'yellow grunsel' or the yellow daffodils to darkness causes flashes that 'produce in our minds sudden radiant ignitions that are vividly mimetic of actually seen light'.[69] The flash of light against the dark canvas of the imagination fades as the lines lead readers across the horizon, but another burst is generated with the description of the scene as 'bronzed with a deep radiance'.[70] In this account of the growth of the poet's mind, Wordsworth does not simply represent cognition as situated but enacts it with a stream of images that engage readers' minds through their bodies.

Thinkers such as Coleridge often misconstrued theories of embodied cognition by assuming that embeddedness entails subjects who are passively subordinate to the environment. Whether this misunderstanding was wilful or genuine, theorists of situated cognition such as Hartley do in fact discuss the active processes by which the mind processes and alters sense data. With poetry, Wordsworth can illustrate these processes both abstractly and morphologically. For example, in the infant-babe passage, his alternating use of active and passive voice makes the mind and the environment function, by turns, as the grammatical subject and then object of the verse, thereby rhetorically demonstrating that they have equal roles in meaning-making. At times, the embodied mind is 'prompt and watchful, eager to combine' the stimuli it encounters.[71] Yet, at others, its 'organs and recipient faculties / *Are* quickened' by stimuli.[72] In the first instance, the infant engages actively with the external world, as it serves as the subject of the action. In the second set of lines, the environment functions as an absent subject that acts on the baby's body. Roused by its limited environment—its mother—the infant's mind enters

[66] Wordsworth, *Prelude*, 1.20.

[67] Wordsworth, *Prelude*, 1.23–26.

[68] Elaine Scarry, *Dreaming by the Book* (Princeton University Press, 2001), p. 164.

[69] Scarry, *Dreaming by the Book*, p. 8.

[70] Wordsworth, *Prelude*, 1.25.

[71] Wordsworth, *Prelude*, 2.247.

[72] Wordsworth, *Prelude*, 2.252–253, emphasis mine.

into a reciprocal relationship with the world in which it is embedded, and the 'feelings' of love that 'pass' between them brings burgeoning awareness 'into his torpid life', awakening the mind, which enables him to make sense of the world.[73] The infant quickly acquires the ability (from its innate capacity) to coalesce amorphous sensory data, shaping into 'one appearance all the elements / And parts of the same object'.[74] These lines illustrate the recursive cycle in which the embodied mind and its environment enact meaning about the world. Wordsworth explicitly states that the infant's mind operates 'in alliance with the works / Which it beholds'.[75] The embodied mind thus plays a fundamental role in Wordsworth's organicism because it organizes the multeity into a meaningful whole.

The Excursion offers another powerful example of the mutually dependent relationship between the human mind and its surround, as it structurally enacts this concept. In the Prospectus, Wordsworth sets out his 'high argument':

> … my voice proclaims
> How exquisitely the individual Mind
> (And the progressive powers perhaps no less
> Of the whole species) to the external World
> Is fitted:—and how exquisitely, too,
> Theme this but little heard of among Men –
> The external World is fitted to the Mind.[76]

The parenthetical asides, along with the repetition of certain words and phrases in this passage, disrupt the flow of meaning, making it difficult to comprehend quickly. The parenthetical—'And the progressive powers perhaps no less / Of the whole species'—contains deliberately and unarguably tortured syntax. In conjunction with the alliteration of the 'p' sounds, the lines are a tongue-twister when read aloud. The inverted grammatical structure in line 67, 'Theme this but little heard of among Men', is even more difficult to recite. Readers who want to understand the basic meaning of these lines, figurative implications aside, must slow down and even stop to parse them. It is impossible for serious readers to glide over these lines.[77] Close examination reveals that

[73] Wordsworth, *Prelude*, 2.244.

[74] Wordsworth, *Prelude*, 2.248–249.

[75] Wordsworth, *Prelude*, 2.274–275.

[76] Wordsworth, *The Excursion*, ed. by Sally Bushell, James A. Butler, and Michael C. Jaye, The Cornell Wordsworth (Cornell University Press, 2007), ll. 71, 63–68, Prospectus.

[77] This passage presents a problem opposite to that which Davidson identifies in the 'Ode', in which 'phrases are simultaneously mysterious yet seemingly understood because readers do not have to struggle with grammatical complexity, nor are brought to a pause in their reading'. Davidson, *The Intelligible Ode*, p. 169.

Wordsworth's semantic and syntactical reversals enact the kind of relationship between the mind and nature that *The Excursion* represents throughout. This passage establishes that neither takes precedence over the other, in contrast to Cartesian dualism, and highlights their reciprocity.

These lines are chiasmatic at every level, from the verbal to the grammatical to the conceptual. Verbally, we can see and hear the repetition of the most important words and phrases that convey the conceptual meaning—'exquisitely', 'Mind', 'external World', and 'is fitted'. The chiasmus hinges on the first appearance of the verb phrase 'is fitted'. If we think of the chiasmus as an 'X' structure, 'is fitted' sits at the centre. In the first arm of the X, the mind functions as the subject, while the world serves as the object. This grammatical construction replicates Cartesian structures in which the human is an active and agential subject, and the inert and lifeless world is the object of perception. Yet Wordsworth flips the construction in the second arm of the X, making the external world the subject with the mind as its direct object. As a sentence, it upsets the Cartesian paradigm by asserting mind–world equivalence syntactically and conceptually. The verb 'fit' communicates a sense of harmony between the mind and the world, and the adverb 'exquisitely' intensifies the effect of mutual dependence, signalling how well they conform to each other. The adverb also gestures towards the body and its exquisite sensitivity (sensibility) to its surround. Thus, as well as establishing the cognitive interchange that underwrites Wordsworth's organicism, these lines grammatically literalize the interdependence of the mind and the world.

In addition to these abstract representations, Wordsworth offers nuanced illustrations of how the embodied mind participates in its own growth. In the skiff-stealing incident of *The Prelude*, for example, the boy's mental and emotional state partially determine the meaning that he makes of the encounter. Wordsworth establishes a detailed scene that, in keeping with his usual pattern, enables his readers to easily imagine it and underscores the importance of sensation in cognition. In this process-based theory of cognition, sensation is not the starting point, for 'Who shall point as with a wand, and say / This portion of the river of my mind / Came from yon foundation?'[78] The linearity of reading requires Wordsworth to depict the process linearly, but there is no identifiable beginning or end point to each formative episode for readers to 'parcel out / … by geometric rules'.[79] Everything happens so fluidly

[78] Wordsworth, *Prelude*, 2.247–249.

[79] Wordsworth, *Prelude*, 2.242–243. Bruhn and Davidson, independently of each other, assert that geometric proofs, which do not rely on sensible evidence, contributed to Wordsworth's transcendental dualism. However, in this passage, which is often read as a critique of empiricist accounts of mind, Wordsworth explicitly considers geometric proofs as unsuited for representing it. I read this passage as a commentary on the difficulty of parsing experience to determine what event or stimulus led to any given belief, habit, or

and quickly that they are functionally inseparable. However, Wordsworth starts with the sensate body as a challenge to Cartesianism. He anchors the setting with concrete particulars, noting that a shepherd owns the 'skiff [that] to a willow-tree was tied / Within a rocky cave', making the scene easy to visualize. He also describes the speaker's situation—a stranger, a schoolboy on holiday—and the mood—'It was an act of stealth / And troubled pleasure'.[80] These details and place markers indicate that the boy does not embark on this adventure *tabula rasa*. In contrast to *The Prelude*'s opening rhetorical question, which draws attention to the boy's embodiment, his environment, but not his mental or emotional state, this scene depicts a mind more developed. His perceptions of himself and what he is doing colour how he experiences and makes sense of it later. As the narrative progresses, Wordsworth continues to foreground sensory details, such as the 'Small circles glittering idly in the moon' and the 'craggy steep'.[81] He also vividly describes the action. The boy's lusty rowing causes the skiff to 'heav[e] through the water like a swan'.[82] These actions highlight the boy's interaction with his environment. Wordsworth points out the bodily effects of the boy's mental and emotional state, drawing attention to his 'trembling hands' as he imagines an angry mountain chasing him.[83] After setting up the phenomenal scene, Wordsworth turns to the mental effects of the incident.

Over the next several days, the boy processes what happened to him as he tries to understand the significance of the event. Although Bruhn sees the meaning that emerges as an act of pure reason, Wordsworth's physiological and psychological language indicates that the process of making meaning is more complex and that the body is implicated beyond perception. Furthermore, the active role that he attributes to nature, the boy's sense that the mountain 'upreared its head' and 'like a living thing, / Strode after' him, points to its role in meaning-making.[84] In enactive accounts of sense-making, organisms attend primarily to details in the environment that they see as salient to themselves. Similarly, the boy's emotional state, his sense of alienation as a 'stranger' in Patterdale, and his belief that his is an 'act of stealth' shapes how the world discloses itself to him.[85] His most vivid recollection is being chased by nature, and as he remembers the event, its 'huge and mighty forms that do

attitude. Thomas Wedgwood, a thoroughgoing materialist, makes a similar comment in his notebooks, noting 'the difficulty ... in investigating the origin of different feelings'. Thomas Wedgwood, 'Correspondence and Papers', V&A Wedgwood Collection, loc. E40 28478.

80 Wordsworth, *Prelude*, 1.373–375, 388–389.

81 Wordsworth, *Prelude*, 1.392, 406.

82 Wordsworth, *Prelude*, 1.401, 404.

83 Wordsworth, *Prelude*, 1.412.

84 Wordsworth, *Prelude*, 1.408, 411–412.

85 Wordsworth, *Prelude*, 1.378, 388.

not live / Like living men moved slowly through [his] mind'.[86] As he thinks about it, 'That spectacle, for many days, [his] brain / Work'd with a dim and undetermin'd sense / Of unknown modes of being'.[87] The reference to the brain suggests that reason is involved, but the 'dim and undetermin'd sense' and 'unknown modes of being' suggest that the unconscious also participates. The meaning at which the boy arrives has mental, emotional, and physical consequences. He believes that the incident purified 'elements of feeling and of thought', and he feels a sense of nature's 'grandeur in the beatings of the heart'.[88] These kinds of quotidian experiences make up the bulk of the poem. Playing with siblings, ice-skating, and other mundane events yield insights that Wordsworth later recognizes as pivotal in his mental development.[89]

Interestingly, the more dramatic events, such as the spots of time, follow the same syntactic and semantic pattern, but they illustrate different cognitive principles. According to Jonathan Wordsworth, M.H. Abrams, and Stephen Gill, the spots of time introduce an 'important new element in *1805*[,] the emphasis on mind as "lord and master"'.[90] However, although he uses this phrase in his definition of spots of time, Wordsworth's representation of these moments undermine this assertion. He describes them as moments in which 'We have had the deepest feeling that the mind / Is lord and master, and that the outward sense / Is but the obedient servant of her will'.[91] The key phrase in these lines is 'deepest feeling'. Having a feeling that the mind has mastery is not the same as saying that it '*Is* lord and master' (emphasis mine). Here, the word 'feeling' functions as an as-if subordinating conjunction that indicates something that seems true even if it is not.[92] Wordsworth represents spots of time as emotionally intense experiences characterized by heightened awareness. In addition to the effect that mental states have on how the world discloses itself, the intensity of the experiences causes them to remain in his

[86] Wordsworth, *Prelude*, 1.424–426.

[87] Wordsworth, *Prelude*, 1.418–420.

[88] Wordsworth, *Prelude*, 1.438, 441.

[89] The snare-raiding incident, which occurs a few lines above this one, follows a similar pattern. Wordsworth sets up a concrete scene with the 'frosty wind', the 'autumnal crocus', 'the smooth hollows', the 'Moon and stars' that light the way. The boy's emotional state, his 'anxious visitation' and sense that he 'seemed to be a trouble to the peace', establish the mood. These give rise to the terrifying impression that nature with 'Low breathings [was] coming after' him. Again, he interprets the incident, when he has had time to think about it, as moral correction. There are many such episodes in *The Prelude* that represent cognition as situated. Wordsworth, *Prelude*, 1.311, 312, 314, 322, 323, 330.

[90] Editorial note by J. Wordsworth, Abrams, and Gill, in Wordsworth, *Prelude*, n. 4, p. 430.

[91] Wordsworth, *Prelude*, 11.270–272.

[92] Another example of this usage in *The Prelude* occurs in Book III, when the narrator states that he had 'a strangeness in my mind / A feeling that I was not for that hour / Nor for that place'. Wordsworth, *Prelude*, 3.79–81.

memory. Their after-effects can be triggered at any time, long after the event has ended, and the percipient is transported mentally or emotionally back in time. Indeed, while the spots of time demonstrate the mind's tenacity, the language that Wordsworth uses to represent them reveals them to be a product of an embodied mind, both at inception and in memory.

In fact, the spots of time illustrate different principles of embodied cognition. For example, the gallows episode details the power of situationally dependent emotions to affect perception. In this scene, the narrator recounts getting separated while out riding with a servant. As a six-year-old boy, he interprets everything as sinister even though later, as an adult, he recognizes that his interpretation was directed by fear. Lost and afraid, he dismounts his horse 'through fear' and wanders around trying to find his companion.[93] When he 'Came to a bottom where in former times / A murderer had been hung in iron chains' and where he can see the man's name inscribed on a monument, his fear intensifies and casts a pall on all that he sees.[94] By the time he encounters the woman carrying a pitcher, he is predisposed to interpret the incident as a scene of struggle and aggravation: the woman 'seemed with difficult steps to force her way / Against the blowing wind' with 'garments vexed and tossed'.[95] Recognizing that his perception of the scene was coloured by his emotional state, the narrator articulates Wordsworth's cognitive theory, stating that 'It was in truth / An ordinary sight', but as he remembers the 'visionary dreariness' with which he invested it, he says that he would 'need / Colours and words that are unknown to man' to describe it. [96] As with the skiff-stealing episode, this incident demonstrates the significant role that mental and emotional states play in how people experience and remember emotionally charged moments.

Wordsworth also illustrates the way emotion, and its long-lasting effects, can become encoded unconsciously. While the death of the boy's father testifies to the mind's power, the mental and emotional states that ensue in later years are triggered by external stimuli and are felt in the body. The event becomes entangled in the affect and the external conditions that the boy felt in the days leading up to the event itself, and, later, similar conditions trigger the same emotions. As he waits peevishly to leave school for the Christmas holiday, he feels 'Feverish, tired, and restless' and watches the road 'impatient for the sight' of the horses that would carry his brothers and him home.[97] Wordsworth's adjectives denote embodied states of being, again implicating the body in the process, while the details of the environment that follow implicate

[93] Wordsworth, *Prelude*, 11.285.

[94] Wordsworth, *Prelude*, 11.288–289.

[95] Wordsworth, *Prelude*, 11.306–307, 314.

[96] Wordsworth, *Prelude*, 11.307–308, 310, 308–309.

[97] Wordsworth, *Prelude*, 11.346–347.

the surround. The narrator recalls that the day was 'Stormy, and rough, and wild', a particular that remains salient in the afteryears. As with the episodes discussed above, the boy construes 'the sorrow' that he feels after his father dies as a 'chastisement', especially when he 'calls to mind / That day so lately past' when he felt irritated waiting for his ride home.[98] This unremarkable incident is inscribed in the boy's memory because it is associated with his father's death. The memory retains an unconscious hold on him well into adulthood, with the power to throw him back to 'That dreary time'.[99] Here, Wordsworth's language implicitly establishes a feedback loop that involves the mind, body, and world. Henceforth:

> … when storm and rain
> Beat on my roof at midnight, or by day
> When I am in the woods, unknown to me
> The workings of my spirit thence are brought.[100]

The mental and emotional state of that 'dreary time' resurfaces under certain environmental conditions, a circumstance that is beyond his conscious control.[101] We can infer that the effect is also physiological, as grief and depression often manifest as a sense of heaviness in the body. The unconscious sway of the past is also discussed by Thomas Wedgwood throughout his private notebooks. Wedgwood observes that he cannot control his physical, mental, and emotional reaction when he hears bells chiming during the day. A keen witness to his own embodied mental processes, Wedgwood traces this unconscious response to the 'striking resemblance between the tone & that of the Hamburg bell with which I had several unpleasant associations & amongst the rest, a sensation of melancholy'.[102] Unable to excise or control the body's response to certain stimuli, the mind is certainly not always 'lord and master' of 'outward sense'.

The final spot of time considered provides a clearcut example of embodied cognition. The narrator's entry into London, which represents sensory overload, does not make sense in any other terms. In contrast to the scene at Mount Snowdon, the imagination does not come to the narrator's rescue. As with the more commonplace events, Wordsworth uses strong visual and kinaesthetic images to describe the young man's arrival in London. As the carriage threads its way through the 'labyrinth of suburban villages', the lines that chronicle

98 Wordsworth, *Prelude*, 11.356, 368–370.

99 Wordsworth, *Prelude*, 11.364.

100 Wordsworth, *Prelude*, 11.385–388.

101 Wordsworth, *Prelude*, 11.364.

102 Wedgwood, V&A Wedgwood Collection E40-28478.

the speaker's first sight of the city are crowded with substantive nouns which underscore his deeply visceral experience.[103] He is surrounded by 'vulgar men ... vulgar forms', 'houses, pavements, streets, of men and things / Mean shapes on every side'.[104] Wordsworth uses asyndeton and repetition to pile up the common objects that overload both the speaker's and the readers' senses. Paradoxically, the speaker interrupts the string of nouns to comment on the experience as it is happening. Exclaiming, '"The threshold now is overpast", great God!', he acknowledges that he is in sensory overload, a state that affects both mind and body.[105] He wonders how stimuli '*external* to the living mind / Should have such sway' over him.[106] The narrator, again paradoxically, experiences overwhelming sensations that feel like he is being crushed physically but at the same time he feels disembodied. 'A weight of ages', he says, sat 'Upon my heart', leaving him with 'no thought embodied, no / Distinct remembrances'.[107] In Wordsworth's poetic lexicon, the heart often functions synecdochally as the body, paired as it is with blood, 'sensations sweet', music, and human beings more generally.[108] The overpowering sights and sounds of the city continue to press upon the speaker, not merely as a weight but as a 'weight *and* power, / Power growing with the weight'.[109] The chiasmus couples the physical sensation of heaviness with immense power. While such power is often associated with the transcendental imagination, here Wordsworth associates it with the embodied mind. Notably, the speaker describes his mind as blank, empty of content, having 'no thought embodied'.[110] Rather than being saved by the transcendent Kantian imagination that steps into sublime moments to assert the mind's ascendancy over the external, he suffers a disturbing sense of disembodiment that deprives him of 'Distinct remembrances'.[111] In the absence of clear memories, one might ask how the speaker is able to describe the scene later. Eugene Gendlin's concept of 'felt sense' helps decipher this seeming inconsistency. Levinson describes Gendlin's research as an exploration of 'the ways in which (and the moments at which) the whole of our experience and

103 Wordsworth, *Prelude*, 8.691.

104 Wordsworth, *Prelude*, 8.695, 696–697.

105 Wordsworth, *Prelude*, 8.700.

106 Wordsworth, *Prelude*, 8.701–702.

107 Wordsworth, *Prelude*, 8.703, 704–705.

108 William Wordsworth, 'Lines Written a Few Miles above Tintern Abbey, on Revisiting the Banks of the Wye During a Tour, July 13, 1798', in *Lyrical Ballads, and Other Poems, 1797–1800 by William Wordsworth*, ed. by James Butler and Karen Green, The Cornell Wordsworth (Cornell University Press, 1992), pp. 116–120 (l. 29). In addition to 'Lines Written Above Tintern Abbey', see also, 'Ode' ('Intimations of Immortality', 1807), 'My Heart Leaps Up' (1807), and 'The Solitary Reaper' (1807).

109 Wordsworth, *Prelude*, 8.705–706, emphasis mine.

110 Wordsworth, *Prelude*, 8.704.

111 Wordsworth, *Prelude*, 8.705.

knowledge is present in a particular sense of confusion or arrest'.[112] Gendlin defines felt sense as the 'body's physical sense of a problem or of some concern or situation' that invites further reflection and, eventually, articulation into language.[113] In this scenario, the body has pre-discursive knowledge that can later be recollected in tranquillity. Each of these spots of time confirms Potkay's contention that *The Prelude* establishes Wordsworth's 'countervailing rejection of mind's ascendency over things', which prevails despite his occasional use of the 'trope of the sovereign mind'.[114]

Even so, it cannot be denied that Wordsworth sometimes represents the imagination in grandiloquent language that makes it seem as if he considers it a transcendent faculty. For example, when he calls it 'an unfathered vapour' in the sixth book of *The Prelude*, he employs an odd metaphor that represents the imagination as independent rather than interdependent, but still phenomenal even if only as a mist.[115] In the Preface to *Poems*, which is his longest and most direct treatment of creative imagination, Wordsworth describes it as a 'sublime consciousness of the soul in her own mighty and divine powers'.[116] He further suggests that it is both immortal and otherworldly. However, he grounds even this seemingly transcendental description of it in the mutual interaction of the body with nature, and his discussion of his creative process reinforces this view. Wordsworth explains that his aim in the first poem under the section headed 'Poems of the Imagination' is to represent 'a commutation and transfer of internal feelings, co-operating with external accidents, to plant, for immortality, images of sound and sight, in the celestial soil of the Imagination'.[117] The poem, 'There Was a Boy', which he later includes in *The Prelude* as the 'Boy of Winander', follows the semantic and syntactic patterns that establish the mind as embodied and embedded. The body takes in the sights and sounds, while the boy's mental and emotional state, his 'internal feelings', set the mood that shapes how he experiences the scene. As in the passage from *The Prelude* depicting the father's death, Wordsworth describes the boy as 'feverish' and full of 'restless anxiety' as he interacts with nature, or 'external accidents'.[118] The word 'commutation', a mathematical and rhetorical

[112] Levinson, *Thinking Through Poetry*, n. 50, p. 251.

[113] Eugene T. Gendlin, *Focusing* (Random House Publishing Group, 1982), p. 80.

[114] Potkay, *Wordsworth's Ethics*, p. 80.

[115] Wordsworth, *Prelude*, 6.527. Jonathan Roberts contends that this 'apostrophe provides a sort of compensation narrative for Wordsworth', having passed over the Alps without noticing that the 'Imagination turns out to be better than reality', an interpretation with which I agree. Jonathan Roberts, 'Wordsworth's Apocalypse', *Literature and Theology*, 20.4 (2006), pp. 361–378 (p. 362).

[116] Wordsworth, 'Preface of 1815', *Prose Works*, III, p. 33.

[117] Wordsworth, 'Preface of 1815', *Prose Works*, III, p. 35.

[118] Wordsworth, 'Preface of 1815', *Prose Works*, III, p. 35. Accidents refer to the perceptible

term, signifies an exchange of two variables and a chiasmus in which the two inverted terms complete each other, respectively. In both definitions, a 'transfer' occurs between the mind, body, and external world that furnishes the imagination with 'life and food / For future years'.[119] Using the metaphor of a garden, which is notably earthy and particularly apt, Wordsworth states that experience plants figurative seeds 'in the celestial soil of the Imagination' that later provides material for poetry and for emotional nourishment.[120] Here, the imagination is more akin to a farmer's field than a disembodied heavenly realm.

The Simplon Pass episode of *The Prelude* is another example in which Wordsworth's high-flown apostrophe to the imagination—'I recognize thy glory'—has been interpreted by critics as a turn to transcendence paradigmatic of displacement, whether of historical repression or of 'the growing independence of Imagination from the immediate external world'.[121] However, read within the framework of situated cognition, this passage positions the imagination as the power that enables humankind to experience itself as connected to the natural world. Furthermore, Wordsworth represents this fundamental principle of organicism—relationality—as phenomenologically knowable irrespective of ontology. However, it must be experienced and understood in terms of embodiment before it can be conceptualized. This passage exemplifies the enactment of knowledge by choregraphing an experience in which readers come to the apocalyptic realization that humankind exists in a co-dependent relationship with nature even as it captures the paradox of organicism.[122] An episode often read as confirming Wordsworth's transcendentalism takes on a significantly different cast when viewed through the lens of embodiment.

In his description of the Gondo Ravine, Wordsworth creates a series of dialectical and oxymoronic visual images that seemingly defy the imagination. The tumult of asyndetic images present a series of contradictory binaries that seem to reinforce the Cartesian paradigm that separates matter from spirit, but a close examination reveals that they can be resolved phenomenologically. After several hours of gloomy trudging, the narrator, disappointed because his

qualities of objects, a term that was adapted from Aristotle and widely used at the time.

119 Wordsworth, 'Tintern Abbey', ll. 65–66.

120 Wordsworth, 'Preface of 1815', *Prose Works*, III, p. 35.

121 Wordsworth, *Prelude*, 6.532; Geoffrey H. Hartman, 'A Poet's Progress: Wordsworth and the "Via Naturaliter Negativa"', *Modern Philology*, 59.3 (1962), pp. 214–224 (p. 219); see also Alan Liu, *Wordsworth: The Sense of History* (Stanford University Press, 1989); Robert A. Brinkley, 'The Incident in the Simplon Pass: A Note on Wordsworth's Revisions', *The Wordsworth Circle*, 12 (1981), pp. 87–102; and Geoffrey H. Hartman, *Wordsworth's Poetry 1787–1814* (Yale University Press, 1971).

122 My use of the term apocalypse 'splits it into an allegorical relationship between textual anticipation and [extra-textual] fulfilment'. Roberts, 'Wordsworth's Apocalypse', p. 361.

unceremonious and oblivious crossing of the Alps failed to live up to expectation, he and his companion encounter:

> ... The immeasurable height
> Of woods decaying, never to be decayed,
> The stationary blast of waterfalls,
> And everywhere along the hollow rent
> Winds thwarting winds, bewildered and forlorn,
> The torrents shooting from the clear blue sky,
> The rocks that muttered close upon our ears—
> Black drizzling crags that spake at the wayside
> As if a voice were in them—the sick sight
> And giddy prospect of the raving stream,
> The unfettered clouds and the region of the heavens,
> Tumult and peace, the darkness and the light ...[123]

As in Percy Shelley's 'Mont Blanc' (1816), Wordsworth crowds the description of the Gondo Ravine with seemingly nonsensical, contradictory binaries. However, these images exemplify how finite nature furnishes humankind with an idea of infinity but without the 'transcendental hook' inherent in idealist models of organicism.[124] The constantly 'decaying' woods that are 'never to be decayed' demonstrate to readers that nature is both ephemeral and eternal. Although the organisms and entities that comprise nature go through the temporal cycle of birth and death, the processes that govern them and of which they are a part are seemingly without end. Even as the individual components that specify nature are continually replaced and renewed, the organic processes persist. Similarly, the 'stationary blast of waterfalls' depicts nature as a powerful and ever-moving force that nonetheless offers its human inhabitants moments of deep stillness. In contrast to the Cartesian thinking that sharply divides mortal from immortal, Wordsworth's portrait of nature in this scene captures the phenomenological sense that nature is not by turns one thing and then its opposite, but is both at the same time. To the human percipient, it is both 'Tumult and peace, the darkness and the light'. The cascading effect of the images additionally creates the condition in which readers enact knowledge. Contrary to Geoffrey Hartman's claim that in this scene 'Wordsworth came to realize that Nature itself led him beyond Nature', nature and the human mind function together and imbue each other with meaning.[125]

[123] Wordsworth, *Prelude*, 6.556–67.

[124] Levinson, *Thinking Through Poetry*, p. 117.

[125] Geoffrey H. Hartman, 'A Poet's Progress: Wordsworth and the *Via Naturaliter Negativa*', *Modern Philology*, 59.3 (1962), pp. 214–224 (p. 214).

Miall argues that Wordsworth's description of the Gondo Ravine 'embodies in their most elemental form the powers that correlate Mind and Nature', as the conflicting images also function as analogies for the mind.[126] For example, the winds that work at cross purposes and the torrents that burst forth from 'the clear blue sky' involve a range of incongruous states that are comparable to human emotions, which are often experienced ambiguously, as 'bewildered' and pointing to many things at once.[127] The 'sense of resistance in the Gondo Ravine, of immense powers in conflict, or barely emergent meaning', Miall contends, 'conveys forces that are operative not only in Nature; it also signifies processes of resistance and conflict that work to form the mind'.[128] The sense that nature and mind function according to similar processes is created by the abstractions that Wordsworth employs in conjunction with concrete sensory language. Furthermore, the grammatical and semantical structures in these lines create the conditions in which readers can enact the same knowledge as the narrator as he reveals it.

Three sentences comprise the twenty-four lines in the Gondo Ravine passage. The first two are short, returning readers to the scene and the mood of 'dull and heavy' disappointment after the exultant apostrophe to the imagination in the previous paragraph.[129] The third sentence, containing seventeen lines, sets up an extended conceit that withholds, until the last four lines, the insight that knowledge emerges from the narrator's engagement with his surroundings. The insight is revealed to the narrator and the readers at the same time. The first part of the sentence situates the narrator and readers in an intensely sensuous environment that leads up to and is necessary to the revelation in the last lines of the paragraph. It consists of a compound subject made up primarily of concrete nouns—'woods', 'waterfalls', 'clouds', 'rocks', and so forth—interspersed with the occasional abstraction—'tumult', 'peace', 'forlorn', 'unfettered'. The two parts of this long sentence are joined by the linking verb 'were', which compares the series of images to mental processes, as Miall notes, but these mental processes are both analogized and enacted when the lines are read. The paradoxical forces of nature, says the narrator,

> Were all like workings of one mind, the features
> Of the same face, blossoms upon one tree,
> Characters of the great apocalypse,
> The types and symbols of eternity,

[126] David S. Miall, 'The Alps Deferred: Wordsworth at the Simplon Pass', *European Romantic Review*, 9.1 (1998), pp. 87–102 (p. 95), doi:10.1080/10509589808570040.

[127] Wordsworth, *Prelude*, 6.561, 560.

[128] Miall, 'The Alps Deferred', p. 96.

[129] Wordsworth, *Prelude*, 6.549.

Of first, and last, and midst, and without end.[130]

These lines introduce the vehicle, 'workings of one mind'. The tenor, in this instance, is doubled; it has two vehicles, the forces of nature ('the stationary blast' and so forth) which comprise the first part of the sentence, but also the human mind, which is implied. They merge in the tenor, thus signalling their mutuality. The vehicle is also compound, using a series of images that combine concrete and abstract nouns that move towards increasing abstraction as they reveal the ideas conveyed by the verse. Significantly, Wordsworth compares the mind and nature using human ('the same face') and natural ('blossoms upon one tree') features. The fevered close of these lines can be understood as the enduring processes of organicism represented in the first half of the sentence. As the contraries resolve into the realization that the mind and nature share the same face and tree, the passage destabilizes its own dualities. It apocalyptically deconstructs the mind/nature binary by revealing that they are both 'types and symbols of eternity', that both will persist as long as the world does. Miall suggests that the '"great apocalypse" perhaps anticipates the union of Man and Nature that would result from the final cathartic resolution of the opposed forces that constitute the "one mind"'.[131] However, the 'great apocalypse' is the interdependence of mind and nature, each of which are its 'characters' (symbols, attributes, avatars), and, therefore, the knowledge that these lines reveal.

The exhilarated tone in these lines not only gestures towards the importance of the revelation but also represents and evokes the transcendent, but still embodied, feeling of being interconnected within a web of relations. George Lakoff and Mark Johnson, early theorists of embodied cognition, account for transcendent experience without making recourse to the noumenal. They explain it as a feeling of kinship with one's milieu, as the sense that we are 'part of *a larger all-encompassing whole*, of *ecstatic participation*—with awe and respect—within that whole, and of the *moral engagement* within such an experience'.[132] Wordsworth's language evokes this rapturous experience. Structurally, linguistically, and figuratively, he enacts the knowledge that neither nature nor mind is an abstraction, and that rather than looking forward to a day when the soul transcends the phenomenal realm, situated embodied interaction with it can lead to ecstatic aesthetic experiences.

[130] Wordsworth, *Prelude*, 6.568–572.

[131] Miall, 'The Alps Deferred', p. 99.

[132] George Lakoff and Mark Johnson, *Philosophy in the Flesh: The Embodied Mind and Its Challenge to Western Thought* (Perseus Books Group, 1999), p. 564.

Ethical Potential

For Lakoff and Johnson, as for Wordsworth, ecstatic feelings that attend aesthetic experiences are not just pleasurable but also activate '*moral engagement*'. Wordsworth saw his organicism as a way to motivate readers to ethical action. His representation of cognition as a feedback loop that involves people and their world aspires to get them to intellectually understand and emotionally experience themselves as part of a web of relations. Furthermore, he believed that, cognitively, people have the capacity 'to intuit (rightly or wrongly) the interconnection of all things'.[133] Wordsworth's situated organicism urges people to recognize and take responsibility for the fundamental relationality that defines their existence. But to do this, they must abandon the Cartesian worldview and see themselves as participants in 'ecological reciprocity in which persons and things commingle and experience life only in their interaction'.[134] While the boy's sense of moral chastisement in *The Prelude* points to ethics, *The Excursion* illustrates that Wordsworth's organicism is no idealist abstraction but obtains as ethical action through the relationships between human beings and their environments.

Through his personal experience and his reading, Wordsworth came to view all of creation as an organic whole. Although he does not discuss his concept of organicism prosaically, expressing his ideas primarily in verse, there are scattered hints in 'Essay, Supplementary' and in the Preface to *Lyrical Ballads* that he explicitly conceived of nature in terms of organicism. Additionally, a number of his ideas about cognition, ethics, society, and nature resemble the theories of embodied organicism articulated by Hartley and Darwin. However, Wordsworth's ideas have the most resonance with Holbach's theories in *System of Nature*. For example, Holbach counsels something like the wise passiveness that Wordsworth advocates in 'Expostulation and Reply' (1798) and 'The Tables Turned' (1798). Holbach also attributes social ills to the 'want of having studied nature and her laws'.[135] Like Holbach, Wordsworth saw human beings' ignorance of their relationship with nature as a source of the problems in society. In *The Excursion* he illustrates the ethical potential of his organicism in the relationships between human beings, nature, and each other. Although many critics, including the second-generation Romantics, saw this poem as a sign of Wordsworth's political quietism, it in fact expresses values similar to those that impelled him to support the French Revolution, particularly his commitment to social justice. Furthermore, it depicts organicism as an

[133] Potkay, *Wordsworth's Ethics*, p. 82.

[134] Jonathan Farina, '*The Excursion* and "The Surfaces of Things"', *The Wordsworth Circle*, 45.2 (2014), pp. 99–105 (p. 100).

[135] Holbach, *System*, I, p. 26.

embodied principle that has the potential to structure people's phenomenological understanding of the world and spur them to ethical action.

While Wordsworth's most explicit prose expressions of organicism in nature are made in context of other topics, it is evident that he saw the natural world similarly to Coleridge, Darwin, and Holbach. For example, in his discussion of James McPherson's Ossian poems in 'Essay, Supplementary' he claims that he could tell that they were forgeries based on McPherson's representation of nature. Attributing this ability to his childhood in the mountains of Northern England, he asserts that based on what he saw as a child 'with my own eyes, I knew the imagery was spurious'.[136] He explains that in 'nature everything is distinct, yet nothing is defined into absolute independent singleness'.[137] In 'McPherson's work, it is exactly the reverse': nature is depicted as 'insulated, dislocated, and deadened,—yet nothing is distinct'.[138] Wordsworth's vision of nature as alive and comprising a whole made up of individual but interconnected parts is not only paradigmatic of organicism but has affinities with Holbach's depiction of it.

In *System*, Holbach theorizes a network of relationships, starting with the smallest unit, the atom, which, like Priestley's architectonic theory of matter, organizes itself into the interconnected structures that comprise nature. However, he goes further than Priestley in that he claims that each organism's structure determines what actions it can take in the world, and that in turn dictates its essence, that is, its identity. According to Holbach, the 'different properties of this matter—its different combinations—its various methods of action, which is the necessary consequence [of its combinations], constitute for us, the *essence* of beings; and it is from these diversified essences, that result the different orders, classes, or systems, that these being occupy, and of which the sum total make, what we call, NATURE'.[139] Holbach's essences are not Platonic or transcendental forms but immanent to each organism or entity, and identity is a function of the system in which organisms are embedded. Holbach notes that 'there can be no independent energy, isolated cause, or detached action in nature'.[140] Wordsworth makes a similar claim in his discussion of knowledge production in the 1836 revision to the Preface to *Lyrical Ballads*. Considering the difference between scientific and poetic knowledge, he insists that nature cannot be understood through isolated facts but only as an interrelated, interdependent whole. He states that 'We have no knowledge, that is, no general principles drawn from the contemplation of

[136] Wordsworth, 'Essay, Supplementary', *Prose Works*, III, p. 77, ll. 578–579.
[137] Wordsworth, 'Essay, Supplementary', *Prose Works*, III, p. 77, ll. 579–580.
[138] Wordsworth, 'Essay, Supplementary', *Prose Works*, III, p. 77, ll. 580–582.
[139] Holbach, *System*, I, p. 31.
[140] Holbach, *System*, I, p. 95.

particular facts'.[141] To make sense of the world, the data gathered by science must be synthesized into a more cohesive understanding of nature. Although brief, these comments reinforce Wordsworth's organic representations of nature in poetry. *The Prelude* and *The Excursion*, however, include humankind in the interdependent system of nature.

For Holbach, as for Wordsworth, the system of nature has implications for human beings, who are as much a part of it as any other species. Humankind's essence, its human nature, is also structurally determined. The higher order cognitive functions that are unique to human beings, for example, the capacity for reason, 'wit, sensibility, imagination, taste, &c. &c.' and all other 'faculties ... called *intellectual* are only certain modes or manners of being and acting, resulting from the organization of our body'.[142] Holbach sees cognition as species specific, which is not to say that he denigrates other animals' cognitive capacities. In contrast to the Cartesian view, he insists that 'it is the very height of folly to refuse the intellectual faculties to animals, they feel—they have ideas—they judge and compare—they choose and deliberate—they have memory—they shew love and hatred'.[143] Yet, only human beings can apprehend the orderly structure of nature and understand it as an ethical imperative because only they can deduce the principles that govern it. Holbach sets out three principles, or laws, that not only explain how nature organizes itself, how it maintains the natural order, and its drive to resist dissolution. These laws also provide a moral map for the species that can perceive them. They are relative co-ordination, the faculty of co-ordering, and conservation. Although Wordsworth does not adopt Holbach's terminology, *The Excursion* illustrates ethics similar to those encoded in these principles, albeit adapted to his own purposes.

For Holbach, relative co-ordination and the faculty of co-ordination are linked, referring to nature as a whole and individuated entities, respectively. The former explains how nature structures itself, including how the organisms and entities that comprise it fit within the system as a whole. Holbach explains that the 'different systems of beings, ... their *particular natures*, depend on the general system of the grand whole of universal nature, of which they form a part, and to which every thing that exists is necessarily attached'.[144] Although

[141] Wordsworth, 'Essay, Supplementary', *Prose Works*, I, p. 140, ll. 404–405.

[142] Holbach, *System*, I, pp. 189, 177.

[143] Holbach, *System*, I, p. 298. This was a common position among materialist-leaning men of science. Hartley, Priestley, and Darwin all thought animals 'differ from us in *degree* only, and not at all in *kind*; having all the same mental, as well as corporeal powers and faculties that we have, though not in the same extent'. Priestley, *Disquisitions*, p. 42.

[144] Holbach, *System*, I, p. 32. Later English editions of *System of Nature*, published in 1820 and 1835, translate 'relative co-ordination' as 'relative adaption' and 'relative adaptation'. Holbach, *System*, I, p. 45; Paul Henri Thiry d'Holbach, *The System of Nature, Or, Laws*

Holbach is considered a mechanist, nature in his view exists in dynamic tension. What Levinson says of Baruch Spinoza applies equally to Holbach; the parts are defined by 'where and how [they] move within the ceaselessly interactive network of ... nature'.[145] The faculty of co-ordering, which complements relative co-ordination, speaks to each organism's capacity to accommodate itself within the system. Defining it somewhat circularly, Holbach states that they have the 'faculty of co-ordering themselves with the whole from which they emanated; without that, they could not be able to subsist'.[146] The difference between relative co-ordination and the faculty of co-ordering is akin to the difference between a noun and a verb. 'Co-ordination' denotes the placement of parts into a relationship of equivalencies, in contrast to subordination. It signifies an orderly arrangement of processes or things, or a state of being in such an arrangement. Co-ordering, on the other hand, indicates the ongoing action of arranging and rearranging so as to maintain proper order. It refers to the capacity of organisms to act, to be acted upon, and to fit, mostly neatly, among the distinct parts of nature. This, too, is a longstanding Wordsworthian belief. For example, in the revised Preface to *Lyrical Ballads*, he asserts that poets understand 'man and nature as essentially adapted to each other, and the mind of man as naturally the mirror of the fairest and most interesting properties of nature'.[147] In the Prospectus to *The Excursion*, he famously states that the mind and the external world are mutually fitted to each other.

Holbach's third principle, conservation, is the propensity of organisms and entities to exercise the faculty of co-ordering. According to Holbach, it is 'the common end towards which all the energies, the force, the faculties of beings seem continually directed'.[148] It can be likened to Spinoza's concept of conatus, the drive to persist. Holbach, however, applies the principle more broadly to include non-living aspects of nature as well. For example, he claims that the 'stone by the strong adhesion of its particles, opposes resistance to its destruction'.[149] Here, 'opposes' means that it effortfully mounts resistance against the forces that would break it down. For Wordsworth, as for Spinoza,

of the Moral and Physical World, trans. by Samuel Wilkinson, 3 vols (P. Davison, 1820), I, p. 91; Paul Henri Thiry d'Holbach, *The System of Nature, Or, Laws of the Moral and Physical World*, trans. by Henry D. Robinson (G.W. & A.J. Matsell, 1835), I, p. 45. Foreshadowing Victorian theories of evolution, these later translations stress organisms' conformation to the environment to ensure survival of the species. The prefix used by Hodgson better indicates the reciprocity that Holbach envisions between the system and its component parts.

145 Levinson, *Thinking Through Poetry*, p. 119.

146 Holbach, *System*, I, p. 145.

147 Wordsworth, 'Preface', *Prose Works*, I, p. 140, ll. 425–427.

148 Holbach, *System*, I, p. 91.

149 Holbach, *System*, I, p. 91. LeBuffe also likens conservation to Thomas Hobbes's notion of self-preservation and Isaac Newton's law of inertia. See LeBuffe, 'Paul-Henri Thiry (Baron) d'Holbach', sec. 3.

the drive towards conservation is not only about survival but also includes the drive to experience 'the deep power of joy'.[150]

In Holbach's view, these three principles of nature have ethical significance that pertain to humankind and the social order. Relative co-ordination is responsible for 'THE ORDER OF THE UNIVERSE', which he juxtaposes against 'CONFUSION'.[151] Holbach equates confusion with chance, which, as a structural determinist, he does not believe occurs in nature. Confusion is not part of the natural order but, rather, reflects human ignorance of nature's laws. Because they have the ability to reason, people have the capacity to perceive and understand the laws of nature, which furnishes knowledge of nature's logic. The 'sight of the necessary periodical and regular motion that takes place in the universe, generates in the minds of men the idea of *order*'.[152] Additionally, it authorizes reason because it 'represents a mode of considering and perceiving with facility together and separately the different relations of the whole'.[153] Put differently, relative co-ordination enables people to understand organicism. Likewise, Wordsworth sees the embodied mind as the lynchpin that enables people to understand the principle of relationality and to act on it. Knowing how the parts and the whole relate to each other is a function of the faculty of co-ordering— once we understand the order of the universe, we can mindfully 'co-order ... ourselves with the beings that environ us or, with the whole of which we make a part'.[154] In other words, once people understand that co-ordination and not subordination is the rule of nature, they can restructure society more equitably, which better facilitates conservation.

Holbach blamed widespread ignorance of nature's laws for the exploitive social structures in pre-revolutionary France and throughout Europe. A staunch atheist, he held the twin powers of Church and State responsible for keeping its citizenry subjugated and in thrall of all sorts of supernatural nonsense. Under the control of 'sovereigns, unjust, incapable, enervated by luxury, corrupted by flattery, depraved by licentiousness and impunity, devoid of talents, of morals and of virtue', the citizenry is a mass of 'poor degraded captive[s], destitute of grandeur of soul, of reason, or of virtue'.[155] The Church enables self-serving monarchs to ill-use the people by making them malleable,

150 Davidson, *The Intelligible Ode*, p. 193. Davidson cogently demonstrates how frequently the word joy appears across Wordsworth's corpus and how much he valued it. See Davidson, passim.

151 Holbach, *System*, I, p. 145.

152 Holbach, *System*, I, p. 101.

153 Holbach, *System*, I, p. 101.

154 Holbach, *System*, I, p. 102.

155 Paul Henri Thiry d'Holbach, *The System of Nature; or, the Laws of the Moral and Physical World*, trans. by William Hodgson, 4 vols (G. Kearsley, 1797), II, p. 505, Eighteenth Century Collections Online <link.gale.com/apps/doc/CW0122703413/

submissive subjects. It accomplishes this by filling their heads with stories about 'monsters', 'wonders', or 'miracles' in Nature that make them superstitious, susceptible to the belief in immaterial and immortal souls, and thus afraid of eternal damnation.[156] Despite the ostensible motive to promote moral behaviours, societies dominated by corrupt governments and religious superstition are rife with 'inveterate hatreds', 'barbarous persecutions', 'continual massacres', and 'dreadful tragedies'.[157] Not only is such a society 'blinded and badly constituted', but it is in 'a state of madness'.[158] People behave wickedly and irrationally because they have not been governed according to their natures, and their lack of knowledge has prevented them from revolting against injustice. Holbach asserts that humankind has been kept 'so long in ignorance' that its progress has been impeded; this is why it has 'taken such slow and irresolute steps to ameliorate [its] condition'.[159] Understanding nature provides people with knowledge of order and of human nature, both of which would help improve society.

The key to a more equitable society, for both Holbach and Wordsworth but to different degrees, is knowing and honouring the laws of nature. Holbach urges humankind to 'study this nature—let him learn her laws—let him contemplate her energies, and the immutable rules by which she acts—let him apply these discoveries to his own felicity'.[160] Wordsworth also held that knowledge of nature could help people both individually and socially. The wise passiveness that he counsels in much of his poetry is similar to Holbach's injunction to 'submit in silence to her [nature's] laws, which nothing can alter'.[161] Notably, Holbach does not recommend emulating nature. As Priestley and Darwin posit, 'One of the greatest and most striking evils in the system of nature, is that one animal should be made to prey upon another'.[162] Nature is 'red in tooth and claw', as Alfred Tennyson famously asserted in 1850, but Holbach believed that the faculty of co-ordering enabled humankind to find its place within the system of nature.[163] He states that 'the aptitude in man to co-order himself with the whole, … not only gives him the idea of order,

ECCO?u=asecs&sid=bookmark-ECCO&xid=a3ec6a55&pg=1> [accessed 15 April 2012]. Holbach, *System*, I, p. xii.

[156] Holbach, *System*, I, p. 108.

[157] Holbach, *System*, I, p. xii.

[158] Holbach, *System*, II, p. 409.

[159] Holbach, *System*, I, pp. 26–27.

[160] Holbach, *System*, I, p. 18.

[161] Holbach, *System*, I, p. 18.

[162] Joseph Priestley, *Institutes of Natural and Revealed Religion*, 2nd edn, 2 vols (Pearson and Rollason for J. Johnson, 1782), II, p. 26, Eighteenth Century Collections Online <https://link.gale.com/apps/doc/CW0122283468/ECCO?sid=bookmark-ECCO&xid=d254db5f&pg=1> [accessed 23 May 2023].

[163] Alfred Tennyson, *In Memoriam*, ed. by Gray, Erik, 2nd edn (W.W. Norton, 2004),

but … also makes him say, that ALL IS RIGHT'.[164] Conforming to the natural order and contributing to the whole as part of the system will restore balance. Indeed, the order of nature provides a blueprint for how to structure society. Human happiness depends on discovering its appropriate place in this system of interrelated parts and acting according to its nature, and when all is properly fitted to the natural order everything functions as it should. Wordsworth makes a similar claim in *The Excursion*, which illustrates the idea that both individuals and societies must abide by the principles of interdependence. The poem reflects Holbach's edict 'never to separate the physical from the moral world', but where he ties morality to the principle of order Wordsworth connects it to relationality.[165]

In Wordsworth's model of society, which mirrors his organicism, understanding how relationality works within a larger system is key to improving social equity. He analogizes society to a set of interconnected parts that function together to produce a whole that surpasses itself. In *The Excursion*, Wordsworth likens society to an organism that follows the rhythms of generation and decay: 'Her organs and members, with decay / Restless, and restless generation, power / and functions dying and produced at need'.[166] The cycle of governments, traditions, and economic systems come and go when society is functioning. However, when there is a serious imbalance it fails and dies, like an organism. Writing to protest against legislation in the 1790s that allowed the wealthy to hoard capital at the expense of the working class, Wordsworth's 'Letter to Bishop Llandaff' compares society to an unhealthy organism that is dying under governmental abuse. He argues that when legislators concentrate power and wealth in one class it threatens society's overall wellbeing because it deprives the working poor of the means to afford even the most basic necessities. A society that fails to distribute its resources is like an organism with a blockage that restricts its vital power to one localized spot. It 'would by its stagnation consign it to barrenness and by its putrefaction infect it with death'.[167] In one sense, he is speaking literally. The 'professed object' of the legislation 'is to lower the price of labour, to compel the workman to be *content* with arbitrary wages evidently too small' for the 'labourer to provide food for himself and his family'.[168] Under such circumstances, this group, which is crucial to the health of the nation, is forced into penury which sinks its members into disease and death, a point that Thomas Beddoes makes in his

56.15, Internet Archive <http://archive.org/details/inmemoriamauthor0000tenn> [accessed 19 January 2024].

[164] Holbach, *System*, I, p. 147.

[165] Holbach, *System*, I, p. 89.

[166] Wordsworth, *Excursion*, 7.1023–1025.

[167] Wordsworth, 'Bishop Llandaff', *Prose Works*, I, p. 48, ll. 647–648.

[168] Wordsworth, 'Bishop Llandaff', *Prose Works*, I, p. 43, ll. 450–452, 499.

pamphlet, *A Guide to Self Preservation and Parental Affection* (1794). Beddoes attributes high child mortality and high rates of illness among the poor to 'want of food or cleanliness, or of attendance when they are sick', a 'loss [that] falls heaviest on the families of hard-working people' who cannot afford the same amenities as the rich.[169] Wordsworth reinforces this point years later in his 1835 postscript to *Yarrow Revisited and Other Poems,* which inveighs against the Poor Law Amendment Act of 1834. Again, he examines the social consequences of forcing people into poverty, which, he argues, drives people 'to despair and death, and to the breach of law by theft or violence'.[170] The degree to which a society functions, like any organism, is contingent on how well the parts work together.

According to Potkay, 'ethics for Wordsworth is about concrete relations, not about abstract theories or principles'.[171] Both *The Prelude* and *The Excursion* offer examples of the very real relationships between the various parts of the phenomenal world. Wordsworth's accounts of situated cognition in the former explore nature's potential to educate human beings in ethics. In *The Excursion,* he considers the social consequences of living according to the principle of relationality. Here, he provides a series of tangible illustrations coupled with elucidatory dialogue. The 'high argument' set out in the Prospectus suggests that society emerges from interaction with nature.[172] In addition to the fitness of the mind and the world to each other, Wordsworth's other theme is

> ... the creation (by no lower name
> Can it be called) which they with blended might
> Accomplish.[173]

The antecedents to the pronoun 'they' are the mind and the world, and 'the creation' that they 'Accomplish' through their 'blended might' is society. The lines that immediately follow inform the reader that, 'Such grateful haunts foregoing', the poem will 'travel near the tribes / And fellowships of men, and see ill sights'.[174] 'Foregoing' can be interpreted in two ways. It can be read as the present participle of forgo, to give up the enjoyment of something—the 'grateful haunts', whether mind or nature. It can also be an adjective that points back to what has gone before, in this case 'the creation' in the preceding lines. Thus, not only does *The Excursion* address social ills, but through this double

[169] Thomas Beddoes, *A Guide for Self Preservation, and Parental Affection*, 3rd edn (J. Johnson and H. Murray, 1794), p. 6.

[170] Wordsworth, 'Postscript, 1835', *Prose Works*, III, p. 241, ll. 51–52.

[171] Potkay, *Wordsworth's Ethics*, pp. 31–32.

[172] Wordsworth, *Excursion*, ll. 71, Prospectus.

[173] Wordsworth, *Excursion*, ll. 69–71, Prospectus.

[174] Wordsworth, *Excursion*, ll. 72–73, Prospectus.

meaning and the juxtaposition of 'might' and 'Accomplish' it casts doubt on the mightiness of such a creation. It is clear that 'might' functions as a noun that signifies the creation's strength and power, yet its enjambment keeps the reader in momentary suspense as it introduces the possibility that it functions as a verb instead. Modally, 'might' expresses uncertainty about the creation produced by the 'blended might' of mind and nature. The compendium of case studies and moral tales that follow the Prospectus supports such a reading. As it explores the relationship between human beings, nature, and society, the poem offers insight into what society might look like, in Wordsworth's view, if people enacted the ideals of organicism.

In *The Excursion*, the Wanderer serves as the exemplar of embodied organicism enacted by human beings. Extending Sally Bushell's claim that 'the opening description of the Wanderer defines him through the context of his profession and the way it physically shapes his life and mind', I trace the Wanderer's context back to his childhood. Related early in the first book, the Wanderer's first years illustrate the development of the his organic sensibilities.[175] His youthful interactions with nature lay the foundation for his character in later life when Wordsworth represents him as a person who recognizes and honours his place within the larger ecosystem of natural objects and people. His childhood 'Among the hills of Athol' in Scotland impress him, and readers, with the importance of both nature and society to human flourishing, while his adult experiences demonstrate the importance of human flourishing to nature and society.[176] Wordsworth's brief account of the Wanderer's youth follows the pattern established in *The Prelude.* He receives an ethical education from his embodied interactions with nature, through experiences both sublime and beautiful. Like the narrator in *The Prelude,* he spends much time alone communing with 'hills', 'stars', and 'woods' during the period in which 'the foundations of his mind were laid'.[177] Wordsworth represents nature's effect on the young Wanderer in tangible terms, stating that 'deep feelings had impressed / Great objects on his mind; with portraiture / And colour so distinct, that on his mind / They lay like substances and almost seemed / To haunt the bodily sense'.[178] Wordsworth's language emphasizes that the mental effects of the child's interactions with nature are not just significant but substantial. And, as with the young man's entry into London, his impressions have weight and heft that have almost bodily effects long after they are over. Additionally, Wordsworth represents nature as agential in these lines as

[175] Sally Bushell, *Re-Reading* The Excursion*: Narrative, Response, and the Wordsworthian Dramatic Voice*, Variation: Nineteenth Century (Ashgate, 2002), p. 225.

[176] Wordsworth, *Excursion*, 1.112.

[177] Wordsworth, *Excursion*, 1.143, 145–146, 132.

[178] Wordsworth, *Excursion*, 1.152–156.

it acts upon the child, impressing him with objects and colour, but the child is also active. The 1845 revision makes this clearer with the added line that these experiences 'Perplexed the bodily sense'.[179] An example of what Levinson identifies as 'Wordsworth's statements of unmediated body-knowledge', this addition suggests that the Wanderer had to make sense of his inchoate bodily sensations by bringing his rational mind to bear on his experiences.[180] Like the boy in *The Prelude*, he participates in meaning-making. In keeping with Potkay's claim that Wordsworth had 'faith in the ethical power of attuning ourselves to … environments', I argue that by contemplating the surround people have the opportunity to develop a phenomenological sense of 'interconnectedness'.[181] In consequence, as an adult, the Wanderer models an expansive view of nonhuman nature: 'Birds and beasts', 'fish', 'reptile[s]', and 'gorgeous insect[s]', 'In his capacious mind, he loved them all'.[182] Not only does he recognize that he lives within a web of relations, but he also respects all parts of the system, both animal and human.

Along with his education in nature, the Wanderer learns the importance of society, in part from his step-father, 'the Teacher of the Village School' that he attended in the winter months.[183] There, he received 'Needful instruction' 'in the lore of right and wrong, the rule / Of human kindness'.[184] An education in nature alone is insufficient because it is only one part of our lifeworld. The Wanderer serves as an exemplar because he is equally connected to the human and natural world. The narrator's description of himself applies equally to the Wanderer (if not more so). He says that he has withdrawn 'from the passions of the world' in order to fix himself 'a still retreat' in nature; he is 'Sheltered, but not to social duties lost, / Secluded, but not buried'.[185] Although the Wanderer leads an itinerant life, he does not reject human society but rather maintains in his later years connections he formed on his rounds as a Pedlar. The Wanderer demonstrates *The Excursion*'s argument that the 'discerning intellect of Man' can know itself only 'When wedded to this goodly universe'.[186] Thus, he illustrates the principles of interdependence and relationality embodied in humankind.

Jonathan Farina observes that with '*The Excursion* Wordsworth turns … to a different model of character than appears in *The Prelude*', that is, 'a model of character dependent on the touch, exchange, and mutability of surfaces' that

179 Wordsworth, *Excursion*, n. to 1.153–156.
180 Levinson, *Thinking Through Poetry*, p. 106.
181 Potkay, *Wordsworth's Ethics*, pp. 204, 71.
182 Wordsworth, *Excursion*, 2.44–47, 249.
183 Wordsworth, *Excursion*, 1.121.
184 Wordsworth, *Excursion*, 1.123, 125–126.
185 Wordsworth, *Excursion*, 5.40, 41–43.
186 Wordsworth, *Excursion*, ll. 52–23, Prospectus.

serve as a 'medium for feeling in-touch with the world outside of oneself'.[187] While the Wanderer is the prime example of this, the stories that the narrator hears and the people that he meets as he and the Wanderer perambulate provide counterexamples. Through them, Wordsworth explores the consequences of losing that in-touch feeling and of neglecting relationships with nature and other people. Those who turn away from social responsibility, society, and nature lose their equanimity. With some variation, these characters suffer the mental agony of discontentment, disappointment, or despair. The Priest provides an example of a woman with 'A mind intolerant of lasting peace' as a result of her 'unrelenting, avaricious thrift'.[188] Although the woman was happy as a child, the first days of her married life were marked by poverty and 'dire dependence' and in response she becomes not just frugal but stingy. As 'her heart / Closed by degrees to charity', she 'sternly hoarded each day's gain'.[189] Even after she achieves financial security, she continues to engage in this behaviour, refusing to help the needy in her community. Wordsworth compares her mental state to 'the agitation of a brook that runs / Down rocky mountains': it makes her start 'With violence' even though she has nothing to fear.[190] The manuscript version makes it explicit that her refusal to give to others is the cause of her fretful mind, stating that the woman's 'hand grew slack in alms-giving'.[191] In addition to this brief anecdote that illustrates the ill-effects of parsimony, Wordsworth provides an extended example in the Solitary.

In contrast to Potkay, who contends that the Solitary's ethical flaw is 'radical social dependency', I argue that it is his rejection of people and withdrawal from community in the face of despondency.[192] The Solitary's life was happy until his wife and his children died. In response to this tragedy, he succumbs to 'An uncomplaining apathy' that leaves him entirely indifferent to 'private interest' or 'public care'.[193] In Holbachian terms, he loses interest in conservation in both the personal and social realms. However, the French Revolution, which promised to renew society by rescuing it from the twin evils of governmental and religious tyranny, inspires new hope in him. His commitment to community is rekindled as the Revolution's 'voice of social transport reached even him!'[194] After it collapsed into the Terror, however, 'he lost / All joy in human nature; was consumed, / And vexed'.[195] He becomes

[187] Farina, '*The Excursion* and "The Surfaces of Things"', p. 99.

[188] Wordsworth, *Excursion*, 6.754, 731.

[189] Wordsworth, *Excursion*, 6.739.

[190] Wordsworth, *Excursion*, 6.756–757, 760.

[191] Wordsworth, *Excursion*, n. to 6.742–745.

[192] Potkay, *Wordsworth's Ethics*, p. 169.

[193] Wordsworth, *Excursion*, 2.218, 221.

[194] Wordsworth, *Excursion*, 2.229.

[195] Wordsworth, *Excursion*, 2.311–313.

so disillusioned that he adopts an itinerant lifestyle, but his motives are very different from the Wanderer's. Rather than splitting his time between nature and people, the Solitary began 'a wandering course of discontent / In foreign lands ... inwardly oppressed / With malady'.[196] He withdraws from society and becomes 'a detached Spectator' instead of participating in life.[197] 'Roaming at large, to observe, and not to feel; / And, therefore, not to act', the Solitary desires neither to move nor to be moved.[198] Even after he settles down, he 'wastes the sad remainder of his hours / In self-indulging spleen' 'at a safe distance from a world / Not moving to his mind'.[199] Refusing to experience emotion or to be affected by the web of relationality that he tries unsuccessfully to escape adds to his misery.

Although Wordsworth's Solitary tries to remain detached and indifferent, with his 'sarcastic smile', his actions belie his attitude.[200] The text reveals that he cannot entirely ignore the ethical claims of his surround. For example, he offers comfort to an orphaned boy with a 'gift' of currants 'by which he strives, / With an intermixture of endearing words, / To soothe the Child, who walked beside him, weeping', calling him 'my little One' and reassuring him that his deceased grandfather is at peace.[201] Later, readers learn of his kindness to a 'homeless Pensioner' who lives on charity, and that the Solitary was moved by him, exclaiming 'I loved the Old Man, for I pitied him!'[202] A foil to the Wanderer, the Solitary illustrates Holbach's claim that ignoring nature, including human nature, has baleful consequences.

The story of Margaret is a more extreme example of the mental anguish caused by retreating from the natural and social worlds. Like the Solitary, her withdrawal is precipitated by tragedy, but her withdrawal is even more devastating because, as a woman, she has fewer opportunities to support herself and her children. Already in poverty, Margaret's husband joins a press gang to earn money for his family. Emotionally and financially lost without her spouse, she slowly deteriorates. She turns away from her responsibilities to nature and to people, particularly her children. As she sinks under the weight of despair, she neglects her garden, which in the summer seemed 'To lag behind the season, and had lost / Its pride of neatness'.[203] Like the Solitary, she takes to roaming, although she cannot wander quite so far. However, when she roams, she

[196] Wordsworth, *Excursion*, 2.321–323.
[197] Wordsworth, *Excursion*, 3.909.
[198] Wordsworth, *Excursion*, 3.900–901.
[199] Wordsworth, *Excursion*, 2.327–328, 331–332.
[200] Wordsworth, *Excursion*, 2.621.
[201] Wordsworth, *Excursion*, 2.531–533, 535.
[202] Wordsworth, *Excursion*, 2.780, 783.
[203] Wordsworth, *Excursion*, 1.754–755.

abandons 'Her solitary infant', leaving him to cry 'aloud' untended.[204] When she finally returns, she tells the Wanderer that as her despair increases, she finds it harder and harder to force herself to return home. Again, like the Solitary, she turns away from the external world with a 'thinking mind / Self-occupied; to which all outward things / Are like an idle matter'.[205] As Margaret's grief becomes all-encompassing, she retreats from human society, and her despair deepens until it consumes her. An example of extreme disconnection, Margaret exemplifies Wordsworth's warning in *The Prelude* against retreating so far into the mind so as to lose touch with the surround: the 'self-created sustenance of a mind / Debarred from Nature's living images / Compelled to be a life unto itself' falls into soul sickness similar to that experienced by *Alastor* in Shelley's 1816 poem.[206] Even though Wordsworth values solitude, he never loses sight of the need to remain connected to and ethically engaged with the world.

Margaret and the Solitary illustrate the consequences of living with a Cartesian worldview. The Wanderer's rebuke to the Solitary reveals the importance of staying grounded in phenomenology: 'having ceased to see / With bodily eyes' those who lose hope in society are 'borne down by love / Of what is lost'.[207] They become disconnected from everything around them, including themselves. Their view of themselves as abandoned and alone shapes how the world discloses itself to them. Retreating into subjectivity, they see and experience themselves as separate, which has devastating consequences for Margaret and her infant. Rather than an ontological fact, relationality emerges as a condition of humankind's *umwelt.* Cartesianism disrupts this way of knowing and relating to the world, thus causing people to minimize, devalue, or even fail to recognize that they are embedded in a milieu. Levinson argues that there is a 'freedom' 'that occurs when we grasp our immersion in the whole of nature. For when we do, we see that nature cannot be said to act on the body. Rather, the body-mind participates actively—literally, mindfully—in the global play of motions and affections'.[208] Conversely, under the Cartesian metaphor we fail to see our immersion in our surround, and, consequently, social structures and laws reflect principles of separation rather than of deep relationality and mutual co-dependence.

In addition to exploring the consequences of disconnection, *The Excursion* contains positive examples of and suggestions for creating a more equitable society. Contrary to Bushell's conclusion that 'the presentation of an ideal social structure represented by a village churchyard at the heart of a small community

[204] Wordsworth, *Excursion*, 1.771.

[205] Wordsworth, *Excursion*, 2.832–834.

[206] Wordsworth, *Prelude*, 6.312–314.

[207] Wordsworth, *Excursion*, 4.170–172.

[208] Levinson, *Thinking Through Poetry*, p. 124.

must be read as a conservative position', I see *The Excursion* as, in part, an anti-capitalist screed that suggests a combination of legal reforms and local interventions.[209] In many ways, it is an extension of Wordsworth's discussion of modernity in the Preface to *Lyrical Ballads*, which condemns industrialization because it blunts the mind of labourers and alienates them from self and others. In *The Excursion*, however, he identifies the central problem of industrial capitalism as its ability to transform everything into use-objects in service of capital. The Wanderer expresses this view in explicit terms:

> ... Our life is turned
> Out of her course, wherever Man is made
> An offering, or a sacrifice, a tool
> Or implement, a passive Thing employed
> As a brute mean, without acknowledgment
> Of common right or interest in the end;
> Used or abused as selfishness may prompt.[210]

Capitalism, in this assessment, violates natural law by taking human lives off course. First, factory work forces human beings out of alignment with the rhythms of nature by disrupting their diurnal cycles. Industrialism requires workers to toil in 'an unnatural light / Prepared for never-resting Labourer's eyes' with shifts that start before dawn and last well past nightfall.[211] Second, it treats human beings as a 'brute mean' for making profit, and, as Wordsworth notes in his 'Letter to Bishop Llandaff', profit comes at the expense of a living wage for labourers. Capitalism violates the ethical potential of organicism because it refuses all parts of the system—the natural world included—the same access to 'common right or interest'. Those with resources hoard them, failing to recognize the larger, long-term effects of the exploitation they enact. They behave as if they are not implicated in a web of relationships, as if they are separate from their environments.

Indeed, through the Solitary, Wordsworth extends his critique of capitalist exploitation to the land. During his life, he witnessed nature becoming a resource for industry and saw that the extraction of raw materials was destroying it. The Solitary says of his time in America that the 'unknit Republic', succumbing 'to the scourge / Of its own [greedy] passions', unhesitatingly sacrifices the land to 'the encroaching axe' with 'dire rapacity'.[212] He accurately diagnoses the effects of settler-colonialism, which bent the landscape

209 Bushell, *Re-Reading* The Excursion, p. 187.
210 Wordsworth, *Excursion*, 9.114–120.
211 Wordsworth, *Excursion*, 8.169–170.
212 Wordsworth, *Excursion*, 3.923–927.

to its will and left massive wreckage in its wake, the structures of which continue to dominate society today. Wordsworth not only critiques America for its abuse of the land, but he also criticizes Britain, which had 'such outrage done to Nature' and 'violated her rights' in pursuit of industry.[213]

Wordsworth recognized that the social imbalances he identified were based in class, and in *The Excursion* he suggests remedies geared towards empowering the poor, employed or otherwise. First, he opposes industry's exploitation of children. Children, says the Wanderer, are 'made / The senseless member[s] of a vast machine / Serving as doth a spindle or a wheel'.[214] In its stead, he recommends compulsory, universal education for all children to provide them with 'intellectual implements and tools'.[215] To be sure, Wordsworth couches this suggestion in the language of ideological indoctrination, albeit ambiguously. He contends that it is the State's 'obligation … to *teach* / Them who are born to serve her and obey'.[216] These lines can be read as acknowledging that the lower orders who are born into social servitude must be taught to obey. Or they can be interpreted to mean that the State has an obligation to educate the classes that have been born into servitude. Neither interpretation is egalitarian, yet some of Wordsworth's motives for proposing universal education were for the benefit of the children rather than the State. Like Mary Wollstonecraft, he saw education as a way of sustaining oneself amidst the drudgery of 'weary life'.[217] Additionally, without education, the poor will remain 'A servile Band among the lordly free!'[218] In this respect, he is similar to Beddoes and other progressives, who 'thought undemocratic societies, in part, resulted from depriving whole groups of people access to information and basic skills that would enable them to resist political oppression and other kinds of exploitation'.[219] While Wordsworth was not as radical as Beddoes, some of his proposals can be read as proto-Marxist.

In *The Excursion* and elsewhere, Wordsworth argues for a greater distribution of wealth and power, more humane systems of relief for the poor, and a pre-Marxist scheme of property ownership. In the Postscript to *Yarrow,*

[213] Wordsworth, *Excursion*, 8.155, 157.

[214] Wordsworth, *Excursion*, 9.161–162.

[215] Wordsworth, *Excursion*, 9.307. Tom Duggett situates Wordsworth's views on education in *The Excursion* in the context of the debate between the Madras and Lancasterian approaches to education, examining his friendship with William Bell. See Tom Duggett, 'The Dramatic End of *The Excursion*', *The Wordsworth Circle* 45.2 (2014), pp. 157–161. Bushell, on the other hand, traces similarities between Adam Smith's views on national education and 'the Wanderer's depiction of the negative effects of industrialisation'. Bushell, *Re-Reading* The Excursion, p. 237.

[216] Wordsworth, *Excursion*, 9.296–297.

[217] Wordsworth, *Excursion*, 9.306.

[218] Wordsworth, *Excursion*, 9.309.

[219] Lisa Ann Robertson, 'Doctor Beddoes's Democratic Toys: Education, Embodiment, and Empowerment', *Essays in Romanticism*, 26.2 (2019), pp. 115–129 (p. 117).

and Other Poems, he argues against the Poor Law Amendment Act, asserting that it is government's duty 'to make such effectual provisions that no one shall be in danger of perishing either through neglect or the harshness of its legislation'.[220] He also objects to the distinction made between the so-called deserving and undeserving poor. The laws extended only two years of relief to persons considered 'able-bodied' enough to work.[221] Wordsworth argues that '*all* persons who cannot find employment ... are entitled to maintenance by law'.[222] Asserting that all members of society deserve access to life's basic necessities in the 1835 Postscript, Wordsworth echoes the Wanderer's pronouncement in *The Excursion* that it is a human 'right / To exist and be supplied with sustenance / And means of life'.[223] In addition to basic human rights, Wordsworth makes the more radical proposal of factories that are owned and operated by the labourers who work there. Such a scheme would enable them to keep more of the profits of their labour and to maintain their jobs when the market fluctuates. He concedes that collectives of labourers would likely tend more towards republicanism than the current 'mixed constitution of government' in Britain.[224] He appeases the wealthy by reassuring them that the 'democratic and republican spirit which [labourers] might be apt to foster would not, however, be dangerous in itself', particularly if 'sufficiently counter-balanced, either by landed proprietorship, or by a Church extending itself so as to embrace an ever-growing and ever-shifting population of mechanics and artisans'.[225] Yet the unstated inverse of Wordsworth's comment is that the increased political and economic power of the working classes would also keep the tyranny of the wealthy classes in check.

In addition to proposing governmental structures of support and the redistribution of factory ownership, Wordsworth also advocated local action. In *The Excursion*, the positive examples of social responsibility demonstrate this idea. For example, the Solitary relates the tale of a 'Housewife' who opens 'Her doors to admit' an elderly dispossessed soldier despite her slender means.[226] She gives him 'coarse but wholesome fare' and a place to live even though he is a stranger.[227] The Solitary states that she did the same thing for him when he first arrived. Responding to the call to duty inherent in relationality, the woman aids those who need assistance and helps to usher them into the community.

[220] Wordsworth, 'Postscript, 1835', *Prose Works*, III, p. 242, ll. 75–77.

[221] W.J.B. Owen and Jane Worthington Smyser, 'Introduction: General to Postscript, 1835', in *The Prose Works of William Wordsworth*, III, pp. 231–237, p. 234.

[222] Wordsworth, 'Postscript, 1835', *Prose Works*, III, p. 240, ll. 27–29.

[223] Wordsworth, *Excursion*, 9.310–312.

[224] Wordsworth, 'Postscript, 1835', *Prose Works*, III, p. 249, ll. 350–351.

[225] Wordsworth, 'Postscript, 1835', *Prose Works*, III, p. 249, ll. 351–356.

[226] Wordsworth, *Excursion*, 2.767, 770.

[227] Wordsworth, *Excursion*, 2.771.

In another example, a couple living in a mountain cottage gives the Wanderer food and shelter when he is lost.[228]

Wordsworth's solutions offer no panacea, yet his embodied organicism, like the rest of the models discussed in this book, offer a way of orienting ourselves towards our world. He also calls us to act. His model of organicism not only urges us to think of and experience ourselves as part of a wide net of relationships but also to respond to the deep ethical imperative inherent in relationality. He encourages us not to give up hope, which he defines as 'the sweet air of futurity', and maintains that to foster hope we must act because 'the food of hope / Is meditated action'.[229] The phrase 'meditated action' captures the sense of mindful engagement that is not a retreat from social problems. My purpose for highlighting Wordsworth's more progressive view is not to claim that he was a radical at heart but to show that, despite his progressive conservativism, his organicism offers ways to consciously engage with social problems, to take local action—no matter how small—to help solve these problems, and to live out (as well as in) hope by honouring the ethical claims of our interconnected web.

[228] Wordsworth, *Excursion*, 5.730–841.

[229] Wordsworth, *Excursion*, 9.25, 20–21.

Chapter Five

Romantic Organicism in the Twenty-First Century

In the Introduction, I promised to revisit a claim that I make repeatedly throughout this book: Romantic models of embodied organicism are relevant in the twenty-first century because they provide a structuring metaphor that invites us, individually and as a society, to evaluate and change the way we think about and orient ourselves towards our lived world. In the story of Romantic organicism, Cartesian dualism plays the villain, and I contend that it remains a villain today. Even though thinkers in Britain, Germany, and France from the eighteenth century and through the Romantic period offered a range of theories in the life sciences, the mind sciences, and aesthetics to replace it, Cartesian dualism has continued to dominate the way we think about our world, other people, and even our own bodies. Rather than acting as if we exist in relationship to everything around us and assuming that our actions have long-term and far-reaching consequences, we instrumentalize nature, people, and ourselves. Samuel Taylor Coleridge and Baron d'Holbach were right to blame René Descartes for drastically altering how people viewed, understood, and related to nature. By positing the natural world as lifeless and mechanistic, substance dualism laid the ground for science to evolve into a form that we recognize today. As F.E. Sutcliffe contends, once humankind is defined as a mind 'separated from a world that is matter, the role of man can only be that of dominating his surroundings'.[1] Furthermore, for that domination to happen, human beings had to be metaphorically separated from the material world to disguise the fact that in actuality everything was being put into service of a relatively small class of people.

[1] F.E. Sutcliffe, 'Introduction', in René Descartes, *Discourse on Method and the Meditations* (Penguin Books Ltd, 1971), p. 21.

In the mid-twentieth century, around the time Gilbert Ryle published *The Concept of Mind* (1949) in which he coined the phrase the 'Ghost in the Machine' to describe the Cartesian subject, substance dualism came under serious attack from both philosophy and science, particularly cognitive science.[2] It was considered 'one of the most reviled philosophical views ever invented'.[3] Yet despite the popularity of cognitive theories that conflate the mind with neuronal activity, thus eliminating the immaterial subject that drove the machine, proponents of 4e cognition have pointed to the mind–body dualism that underpins them. Even post-structuralism, which promised to do away with the unified, agential subject altogether, reinscribes the transcendentalist abstractions it challenges. Arguing that identity is discursively constructed, it makes 'the connection between liberation and anti-essentialism … in strikingly immaterial terms' by positing an abstract human independent from biological constraints.[4] Indeed, the 'increasing interest in the problem that "consciousness" and "mind" poses to the essentially Cartesian world-view of natural science' lends credence to the claim that this worldview persists.[5]

In my estimation, the emergence of 4e theories of cognition over the past twenty-five or so years indicates a desire to change the dominant ways of understanding and relating to the world. In expanding the definition of cognition so that it applies to even the most modest of nonhuman species and, in some cases, granting emotions and agency to these structurally simple organisms, 4e cognitive theory challenges species chauvinism. As such it offers robust support to posthuman theory.[6] Furthermore, it reduces the distance between the subject and the object, destabilizes the binary of knower/known, and emphasizes relationality, interdependence, and contingency. Like Romantic theories of embodied organicism, it offers an alternative philosophical model for thinking about how we orient ourselves towards the world we inhabit and a potential corrective to the pernicious effects of Cartesianism.

The uncanny parallels between 4e accounts of cognition and theories of organicism raise questions about the historical conditions that impelled thinkers in both periods to advance theories of relationality, grounded in cognition no less, to challenge Cartesian dualism. For example, why does

[2] Gilbert Ryle, *The Concept of Mind* (Penguin Books, 1949), p. 23.

[3] Mark Rowlands, *The New Science of the Mind: From Extended Mind to Embodied Phenomenology* (Bradford Books, 2010), p. 12.

[4] Donald R. Wehrs, 'Epilogue: Literary Theory and Cognitive Studies', in *Cognition, Literature, and History*, ed. by Mark J. Bruhn and Donald R. Wehrs (Routledge, 2014), pp. 243–252 (p. 246).

[5] C.U.M. Smith, 'Coleridge's "Theory of Life"', *Journal of the History of Biology*, 32.1 (1999), pp. 31–50 (p. 38–46).

[6] See Ralph D. Ellis, *Curious Emotions: Roots of Consciousness in Motivated Action* (John Benjamins Publishing, 2005).

resistance to Cartesianism emerge at these points in history, and why do the solutions theorize mind–body–world interdependence? Of course, the Cartesian paradigm is not the same now as it was in the Romantic period. In the eighteenth and nineteenth centuries, science was moving towards embodiment as it examined physiology, cognition, and life. Comparative anatomy, one of the strongest arguments against the Cartesian distinction between humankind and all other lifeforms, along with observations of animal and insect behaviour, led thinkers such as David Hartley, Joseph Priestley, and Erasmus Darwin to argue that animals and plants differed from human beings only in degree, not in kind. Additionally, the account of nature advanced by Cartesian physics was so phenomenologically inaccurate that thinkers were driven to find better ways to account for it. At the same time, however, the rise of industry exploited bodies by making them work long, untenable hours, operate dangerous machines, and subsist on inadequate wages. Factory work increasingly alienated labourers from product. The theories of organicism advanced during this period not only emphasized relationality, but they also used aesthetics to encourage people to see and experience their world as a web of relationships creating ethical obligations.

In contrast to the Romantic period, today science and technology appear to be moving towards disembodiment.[7] Virtual reality chatrooms, for example, enable users to interact with each other via avatars. These avatars are not just a face in the corner of the screen but have full 'bodies' that mirror users' motions. Furthermore, they often do not even take human form. Users can be bunnies, Manga characters, a banana, or dalek-style robots (to name just a few). Additionally, users can select different 'worlds' to visit, all for a price, of course. At the same time, bodies continue to be exploited on a global scale, forced to engage in activities that destroy them over time. Where Romantic organicism responded to scientific interest in the body by embracing it, perhaps 4e theories of cognition are responding to the drive towards disembodiment by recuperating the body and its surround. It is no coincidence 4e theories were initially inspired by the cognitive theory that grew out of the 1940s–50s Macy Conferences on Cybernetics. After all, in addition to cognitive science, these conferences birthed research into computers that has culminated in virtual reality headsets and large language model chatbots that can pass the Turing test. Perhaps the most notable similarity between the two periods is the exploitation of bodies—human and nonhuman—and nature in service of capital.

Another important difference is that 4e theories of cognition, while they

[7] For the canonical discussion about the illusory disembodiment of data, see N. Katherine Hayles, *How We Became Posthuman: Virtual Bodies in Cybernetics, Literature, and Informatics* (University of Chicago Press, 1999). Hayles's earlier article is also useful; see 'The Materiality of Informatics', *Configurations*, 1.1 (1993), pp. 147–170.

bear strong resemblances to Romantic theories of organicism, do not have an ethical component. Nor do individual theorists seem to see these relational models as having larger ramifications beyond cognitive science. There appears to be little interest, if any, in these paradigms' potential as a metaphor that could restructure our collective worldview. Furthermore, while some theorists have shown an interest in understanding aesthetic experiences in terms of 4e models of cognition, not many consider their transformative potential.[8]

These missing components, I argue, signal what we can learn from Romantic models of embodied organicism, most especially their belief in the efficacy of aesthetics to effect change. Poetry's potential lies in the opportunities it offers for abstract and critical thinking irrespective of ideology. According to Robert Kaufman, the 'formal processes that produce—in the mind's engagement with the dynamics, textures, and resistances of art—critical thinking itself is a form of truth … [that] does not necessarily translate into a progressive politics, but is nonetheless essential to it'.[9] Similarly, I contend that Romantic structures of engagement enable embodied aesthetic experiences that can, but are not guaranteed to, lead to transformative embodied knowledge about the world and oneself that can change behaviour. Miranda Anderson, Michael Wheeler, and Mark Sprevak contend that the 'realisation of the distributed nature of cognition' not only 'upholds the role of the humanities in wider society', but it also 'more broadly challenges humans' ways of being in the world'.[10] Such challenges present opportunities for change, particularly in pedagogical situations such as the university classroom.

Although Jerome McGann maintains that in shifting his focus from fear to hope Wordsworth 'lost the world merely to gain his own immortal soul' and that 'Shelleyan Hope is the reflex of his Idealism', I argue that the hope offered by Romanticism is a 'politics of the future'.[11] Jerome Christensen stakes a claim for Romanticism in 'The Romantic Movement at the End of History' (1994), which he published in response to Francis

[8] For an example of researchers who study aesthetic experience, see Mia Burnett and Shaun Gallagher, '4E Cognition and the Spectrum of Aesthetic Experience', *JoLMA*, 1.2 (2020), pp. 157–175, doi:10.30687/Jolma/2723-9640/2020/02/001. For an example of a researcher who discusses art's affective potential, see Ellis, 'Learning about Emotion Through the Arts', in *Curious Emotions*, pp. 167–187.

[9] Robert Kaufman, 'Legislators of the Post-Everything World: Shelley's "Defence of Adorno"', *ELH*, 63.3 (1996), pp. 707–733 (p. 709).

[10] Miranda Anderson, Michael Wheeler, and Mark Sprevak, 'Distributed Cognition and the Humanities', in *Distributed Cognition in Enlightenment and Romantic Culture*, ed. by Miranda Anderson, George Rousseau, and Michael Wheeler, The Edinburgh History of Distributed Cognition, 4 vols (University of Edinburgh Press, 2019), III, pp. 1–17 (p. 17).

[11] Jerome McGann, *The Romantic Ideology: A Critical Investigation* (University of Chicago Press, 1983), pp. 88, 127; Jerome Christensen, 'The Romantic Movement at the End of History', *Critical Inquiry*, 20.3 (1994), pp. 452–476 (p. 475).

Fukuyama's claim that history had ended when liberalism burked Marxism by absorbing conservatism and naturalizing change as inevitable rather than radical. Fukuyama makes the defeatist claim that 'the endpoint of mankind's ideological evolution and the universalization of Western liberal democracy [is] the final form of human government'.[12] Also responding to Fredric Jameson's injunction to 'Always historicize!', Christensen contends that epochalist narratives of human progress reify injustice as a natural part of society because, in this strain of neo-Marxism, it will persist until history ends.[13] Regarding the Marxist tenet that history is progressive, he observes that 'despite history's progress nothing fundamental has changed'.[14] As an antidote, Christensen offers Romantic hope. He argues that Romantic resistance was 'practiced without good reason in the aftermath of utopian dreams' and therefore 'might be called a politics of hope'.[15] The Reign of Terror, the Napoleonic wars, Peterloo all signalled the death of the dream promised by the French Revolution. Yet people living in the Romantic period continued to resist even when there was no good reason to do so. Christensen defines Romanticism as the 'real movement of feeling that challenges the present stage of things' by anachronistically holding on to the desire for justice and social equity that can only be fulfilled sometime in the future.[16] I argue that the 'movement of feeling' occurs in the body and is recruited by Romantic structures of engagement.

The situation today is grim, even hopeless in ways that I do not need to rehearse, but, according to Christensen, Romantic hope is the 'politics of the future and always will be until something better comes along'.[17] I argue that embodied organicism and Romantic aesthetics have much to offer in bleak political moments because they can help us feel a sense of connection that reminds us that we are not isolated and disconnected, as the Cartesian metaphor implies, and that we exist in relationship whether we acknowledge it or not. Indeed, acknowledging our connections enables us to take actions that can incrementally work towards the 'politics of the future'. In an attempt to reinvigorate vitiated cultural materialism, Brian Massumi argues that 'even if the everyday' is not a 'place of rupture and revolt … it still might be a site for modest acts of "resistance" or "subversion" keeping alive the possibility

[12] Francis Fukuyama, 'The End of History?', *The National Interest*, 16 (1989), pp. 3–18 (p. 4).

[13] Fredric Jameson, *The Political Unconscious: Narrative as a Socially Symbolic Act* (Cornell University Press, 1983), p. 9.

[14] Christensen, 'The Romantic Movement', p. 464.

[15] Christensen, 'The Romantic Movement', p. 471.

[16] Christensen, 'The Romantic Movement', p. 457.

[17] Christensen, 'The Romantic Movement', p. 475.

of systemic change'.[18] He clarifies that this can be accomplished only if we return to 'Matter, movement, body, sensation', which, from the perspective of embodied organicism, is precisely what aesthetics can enable.[19] I contend that embodied aesthetics offers a modest way to resist the pressures of the everyday because its structures rely on the moving, sensing, feeling body. Aesthetic experiences make real the promise of Romantic organicism by enabling us to experience new forms of thinking and new ways of being in the world. Together with 4e theories of cognition, embodied organicism enables us to envision interdependence between all the parts of a system and, in consequence, take small, local steps to counter the damage wreaked by the Cartesian metaphor of separation.

Acting within one's community and surround is, perhaps, more viable because the concrete environment in which we are embedded is conceptually easier to grasp than the abstractions of global citizenship and *the* environment. Of course, we should do whatever we can to combat global injustice and environmental destruction, which often go hand in hand, but at the same time resisting and making local interventions holds out hope for the politics of the future. Although embodied organicism cannot solve the world's problems, it calls people to see themselves in relationship to rather than separate from their bodies, each other, and their natural home and to engage actively and mindfully with them, and, hopefully, to improve their small corners of the world.

[18] Brian Massumi, *Parables for the Virtual: Movement, Affect, Sensation* (Duke University Press, 2002), p. 2.

[19] Massumi, *Parables for the Virtual*, p. 4.

Bibliography

Abrams, M.H., *The Correspondent Breeze: Essays on English Romanticism* (W.W. Norton, 1984)

——, *The Mirror and the Lamp: Romantic Theory and the Critical Tradition* (Oxford University Press, 1971)

Alaimo, Stacy, 'Trans-Corporeal Feminisms and the Ethical Space of Nature', in *Material Feminisms*, ed. by Susan Hekman and Stacy Alaimo (Indiana University Press, 2008), pp. 237–264

Allen, Richard C., *David Hartley on Human Nature* (State University of New York, 1999)

Altman, Matthew C., ed., *The Palgrave Handbook of German Idealism*, Palgrave Handbooks (Palgrave Macmillan, 2014)

Anderson, Miranda, *The Renaissance Extended Mind* (Palgrave Macmillan, 2015)

Anderson, Miranda, Michael Wheeler, and Mark Sprevak, 'Distributed Cognition and the Humanities', in *Distributed Cognition in Enlightenment and Romantic Culture*, ed. by Miranda Anderson, George Rousseau, and Michael Wheeler, The Edinburgh History of Distributed Cognition, 4 vols (University of Edinburgh Press, 2019), III, pp. 1–17

Armstrong, Charles I., *Romantic Organicism: From Idealist Origins to Ambivalent Afterlife* (Palgrave Macmillan, 2003)

Azarias, Brother, *Aristotle and the Christian Church* (Kegan Paul, Trench, and Co., 1888)

Barker-Benfield, G.J., *The Culture of Sensibility: Sex and Society in Eighteenth-Century Britain* (University of Chicago Press, 1992)

Barr, Mark L., 'The Lyric Dispensation: Coleridge, Mosaic Law, and Equivocal Authority in "The Eolian Harp"', *Studies in Romanticism*, 44.3 (2005), pp. 293–316

Beatty, Arthur, *William Wordsworth: His Doctrine and Art in Their Historical Relations*, 2nd edn (University of Wisconsin, 1927)

Becker, Gavin de, *The Gift of Fear: Survival Signals That Protect Us from Violence* (Dell, 1998)

——, *Protecting the Gift: Keeping Children and Teenagers Safe*, reprint edn (Dell, 2000)

Beddoes, Thomas, *A Guide for Self Preservation, and Parental Affection*, 3rd edn (J. Johnson and H. Murray, 1794)

Bennett, Jane, *Vibrant Matter: A Political Ecology of Things* (Duke University Press, 2010)

Beretta, Alan, 'Joseph Priestley: An Instructive Eighteenth Century Perspective on the Mind-Body Problem', in *Brain, Mind and Consciousness in the History of Neuroscience*, ed. by C.U.M. Smith and Harry A. Whitaker, History, Philosophy and Theory of the Life Sciences, 6 (Springer, 2007), pp. 75–89

Berkeley, Richard, 'The Providential Wreck: Coleridge and Spinoza's Metaphysics', *European Romantic Review*, 17.4 (2006), pp. 457–475

Bewell, Alan, 'Erasmus Darwin's Cosmopolitan Nature', *ELH*, 76.1 (2009), pp. 19–48

——, *Wordsworth and the Enlightenment: Nature, Man, and Society in the Experimental Poetry* (Yale University Press, 1989)

Beyer, Christian, 'Edmund Husserl', in *The Stanford Encyclopedia of Philosophy*, ed. by Edward N. Zalta and Uri Nodelman, Summer 2018 (Metaphysics Research Lab, Stanford University) <https://plato.stanford.edu/archives/sum2018/entries/husserl/> [accessed 5 December 2019]

Blackburn, William, 'Carlyle and the Composition of "The Life of John Sterling"', *Studies in Philology*, 44.4 (1947), pp. 672–687

Blake, William, *The Marriage of Heaven and Hell* (Oxford University Press, 1975)

Boulanger, Erin F., and others, 'Sugar-Phosphate Toxicities', *Microbiology and Molecular Biology Reviews*, 85.4 (2021), pp. 1–28

Bowen, Ralph H., 'Preface', in *Rameau's Nephew and Other Works*, by Denis Diderot, trans. by Jacques Barzun and Ralph H. Bowen (Macmillan Library of Liberal Arts, 1964), pp. 89–91

Brinkley, Robert A., 'The Incident in the Simplon Pass: A Note on Wordsworth's Revisions', *The Wordsworth Circle*, 12 (1981), pp. 87–102

Brown, Thomas, *Observations on the Zoonomia of Erasmus Darwin, M.D.* (J. Johnson and J. Wright for Mondell & Son, 1798), Eighteenth Century Collections Online <link.gale.com/apps/doc/CW0106965553/ECCO?u=asecs&sid=bookmark-ECCO&xid=70b022cf&pg=1> [accessed 6 November 2008]

Bruhn, Mark J., '*The Prelude* as a Philosophical Poem', in *The Oxford Handbook of William Wordsworth*, ed. by Richard Gravil and Daniel Robinson, Oxford Handbooks (Oxford University Press, 2015), pp. 397–413

——, *Wordsworth Before Coleridge: The Growth of the Poet's Philosophical Mind, 1785–1797* (Routledge, 2018)

Budge, Gavin, 'Erasmus Darwin and the Poetics of William Wordsworth: "Excitement without the Application of Gross and Violent Stimulants"', *Journal for Eighteenth-Century Studies*, 30.2 (2007), pp. 279–308, doi:10.1111/j.1754-0208.2007.tb00337.x

——, *Romanticism, Medicine and the Natural Supernatural: Transcendent Vision and Bodily Spectres, 1789–1852* (Palgrave Macmillan, 2012)

Buell, Lawrence, *The Environmental Imagination: Thoreau, Nature Writing, and the Formation of American Culture* (Belknap Press of Harvard University Press, 1995)

Burke, Edmund, *A Philosophical Enquiry into the Sublime and Beautiful*, ed. by Paul Guyer, Oxford World's Classics (Oxford University Press, 2015)

Burnett, Mia, and Shaun Gallagher, '4E Cognition and the Spectrum of Aesthetic Experience', *JoLMA*, 1.2 (2020), pp. 157–175, doi:10.30687/Jolma/2723-9640/2020/02/001

Bushell, Sally, *Re-Reading* The Excursion*: Narrative, Response, and the Wordsworthian Dramatic Voice*, Variation: Nineteenth Century (Ashgate, 2002)

Butler, Joseph, *The Analogy of Religion, Natural and Revealed, to the Constitution and Course of Nature* (James, John and Paul Knapton, 1736), Eighteenth Century Collections Online <https://link.gale.com/apps/doc/CW0118843279/ECCO?sid=bookmark-ECCO&xid=bd928775&pg=5> [accessed 22 April 2024]

——, *Fifteen Sermons Preached at the Rolls Chapel* (Botham, 1726)

Butler, Judith, 'Sex and Gender in Simone de Beauvoir's *Second Sex*', *Yale French Studies*, 72 (1986), pp. 35–49

Cajori, Florian, *A History of Mathematical Notations* (Dover Publications, 1993)

Carey, Daniel, 'Francis Hutcheson's Philosophy and the Scottish Enlightenment', in *Scottish Philosophy in the Eighteenth Century: Morals, Politics, Art, Religion*, ed. by Aaron Garrett and James A. Harris, A History of Scottish Philosophy, 2 vols (Oxford University Press, 2015), I, pp. 36–76

Carlyle, Thomas, *The Life of John Sterling* (Chapman and Hall, 1871)

Cheyne, Peter, '"Wild Activity of Thoughts": Associationism and the Idea of Life', *Coleridge Bulletin: The Journal of the Friends of Coleridge*, n. s., 52 (2018), pp. 17–26

Christensen, Jerome, *Coleridge's Blessed Machine of Language* (Cornell University Press, 1981)

——, 'The Romantic Movement at the End of History', *Critical Inquiry*, 20.3 (1994), pp. 452–476

Christie, Will, 'Wordsworth and the Language of Nature', *Wordsworth Circle*, 14.1 (1983), pp. 40–47

Clark, Andy, and David Chalmers, 'The Extended Mind', *Analysis*, 58.1 (1998), pp. 7–19

Coburn, Kathleen, *Inquiring Spirit: A Coleridge Reader* (Minerva Press, 1951)

Cockin, William, *The Freedom of Human Action Explained and Vindicated: In Which the Opinions of Dr. Priestley on the Subject Are Particularly Considered* ([n. pub.], 1791)

Coleridge, Samuel Taylor, *Biographia Literaria, or, Biographical Sketches of My Literary Life and Opinions*, ed. by James Engell and W. Jackson Bate, Collected Works of Samuel Taylor Coleridge, 7, 2 vols (Princeton University Press, 1983)

——, 'The Eolian Harp: Composed at Clevedon, Somersetshire', in *Poetical Works*, ed. by J.C.C. Mays, The Collected Works of Samuel Taylor Coleridge, 16, 2 vols (Princeton University Press, 2001), I, pp. 231–235

——, *The Friend*, ed. by Barbara E. Rooke, The Collected Works of Samuel Taylor Coleridge, 4, 2 vols (Princeton University Press, 1969)

——, *Lectures 1818–1819: On the History of Philosophy*, ed. by J. R. de J. Jackson, The Collected Works of Samuel Taylor Coleridge, 8, 2 vols (Princeton University Press, 2000)

——, *Marginalia*, ed. by George Whalley and H.J. Jackson, The Collected Works of Samuel Taylor Coleridge, 12 (Princeton University Press, 1980)

——, 'Note on Spinoza', *The Collected Works of Samuel Taylor Coleridge: Shorter Works and Fragments*, ed. by H.J. Jackson and J.R. de J. Jackson, Collected Works of Samuel Taylor Coleridge, 11, 2 vols (Princeton University Press, 1995), I, pp. 607–624

——, *The Notebooks of Samuel Taylor Coleridge*, ed. by Kathleen Coburn, 5 vols (Routledge & Kegan Paul, 1957), I

——, *The Notebooks of Samuel Taylor Coleridge*, ed. by Kathleen Coburn, 5 vols (Routledge & Kegan Paul, 1961), II

——, *The Notebooks of Samuel Taylor Coleridge*, ed. by Kathleen Coburn, 5 vols (Routledge & Kegan Paul, 1973), III

——, *Table Talk*, ed. by Carl Woodring, The Collected Works of Samuel Taylor Coleridge, 14, 2 vols (Princeton University Press, 1990)

——, 'Theory of Life', *The Collected Works of Samuel Taylor Coleridge: Shorter Works and Fragments*, ed. by H.J. Jackson and J. R. de J. Jackson, Collected Works of Samuel Taylor Coleridge, 11, 2 vols (Princeton University Press, 1995), I, pp. 481–557

Craig, Cairns, '"Kant Has Not Answered Hume": Empiricism and the Romantic Imagination', in *Romantic Empiricism: Poetics and the Philosophy of Common Sense, 1780–1830*, ed. by Gavin Budge, The Bucknell Studies in Eighteenth-Century Literature and Culture (Bucknell University Press, 2007), pp. 40–63

Crocker, Lester B., *Diderot's Chaotic Order: Approach to Synthesis* (Princeton University Press, 1974)

Damasio, Antonio, *Looking for Spinoza: Joy, Sorrow, and the Feeling Brain* (Harcourt, 2003)

Darwin, Erasmus, *The Botanic Garden: A Poem in Two Parts. Part I. Containing The Economy of Vegetation. A Poem. With Philosophical Notes*, 3rd edn (J. Johnson, 1795)

——, *The Botanic Garden: A Poem in Two Parts. Part II. Containing The Loves of the Plants. A Poem. With Philosophical Notes*, 3rd edn (J. Johnson, 1795)

——, *The Temple of Nature: Or, The Origin of Society: A Poem, with Philosophical Notes* (J. Johnson, 1803)

——, *Zoonomia or, the Laws of Organic Life*, Cambridge Library Collection, 2 vols (Cambridge University Press, 2009)

——, *Zoonomia; Or, The Laws of Organic Life*, 3rd edn, 4 vols (J. Johnson, 1801)

Davidson, Graham, *The Intelligible Ode: Intimations of Paradise* (The Lutterworth Press, 2023)

Davy, Humphry, 'MS. Notebooks' (before 1800), The Royal Institution of Great Britain.

Deleuze, Gilles, *Spinoza: Practical Philosophy*, trans. by Robert Hurley (City Lights Books, 1988)

Dempsey, Liam P., 'Thinking-Matter Then and Now: The Evolution of Mind-Body Dualism', *History of Philosophy Quarterly*, 26.1 (2009), pp. 43–61

Derrida, Jacques, *Limited Inc*, ed. by Gerald Graff, trans. by Jeffrey Mehlman and Samuel Weber (Northwestern University Press, 1988)

Descartes, René, *Discourse on Method and the Meditations*, trans. by F.E. Sutcliffe (Penguin Books, 1971)

——, *The Philosophical Writings of Descartes*, trans. by John Cottingham, Robert Stoothoff, and Dugald Murdoch, 2 vols (Cambridge University Press, 1984)

Diderot, Denis, *Diderot, Interpreter of Nature: Selected Writings*, ed. by Jonathan Kemp, trans. by Jean Stewart and Jonathan Kemp, 2nd edn (International Publishers, 1963)

Dixon, Thomas, *From Passions to Emotions: The Creation of a Secular Psychological Category* (Cambridge University Press, 2003)

Duggett, Tom, 'The Dramatic End of *The Excursion*', *The Wordsworth Circle* 45.2 (2014), pp. 157–161

Durant, Graham, and W. Ian Rolfe, 'William Hunter (1718–1783) As Natural Historian: His "Geological" Interests', ed. by Hugh Torrens, *Earth Sciences History*, 3.1 (2007), pp. 9–24, doi:10.17704/eshi.3.1.6451t601508g3541

Dushane, Allison, '"Mere Matter:" Causality, Subjectivity and Aesthetic Form in Erasmus Darwin', *Romanticism and Victorianism on the Net*, 56 (2009), doi:10.7202/1001100ar

Eddy, Matthew Daniel, 'Geology, Minerology and Time in John Walker's University of Edinburgh Natural History Lectures (1779–1803)', *History of Science*, 39 (2001), pp. 95–119

Ellermann, Greg, 'Late Coleridge and the Life of Idealism', *Studies in Romanticism*, 54.1 (2015), pp. 33–56

Ellis, Ralph D., *Curious Emotions: Roots of Consciousness in Motivated Action* (John Benjamins Publishing, 2005)

Engelstein, Stefani, *Anxious Anatomy: The Conception of the Human Form in Literary and Naturalist Discourse*, Studies in the Long Nineteenth Century (State University of New York Press, 2008)

Esposito, Stefan, 'Communicating Life: Mary Shelley's *Frankenstein* and Romantic Organicism', in *Mary Shelley: Her Circle and Her Contemporaries*, ed. by L. Adam Mekler and Lucy Morrison (Cambridge Scholars, 2010), pp. 27–43

Fairchild, Hoxie N., 'Hartley, Pistorius, and Coleridge', *PMLA*, 62.4 (1947), pp. 1010–1021

Fairer, David, *Organising Poetry: The Coleridge Circle, 1790–1798* (Oxford University Press, 2009)

Farina, Jonathan, '*The Excursion* and "The Surfaces of Things"', *The Wordsworth Circle*, 45.2 (2014), pp. 99–105

Favela, Luis H., 'Dynamical Systems Theory in Cognitive Science and Neuroscience', *Philosophy Compass*, 15.8 (2020), pp. 1–5, doi:10.1111/phc3.12695

Felluga, Dino Franco, *Critical Theory: The Key Concepts*, Routledge Key Guides (Routledge, 2015)

Foca, Anna, '"Let Me Be the Calm You Seek": Imagination as (Safe)House in Wordsworth's "Tintern Abbey"', *Bucknell Review*, 42.2 (1998), pp. 31–42

Fowler, Thomas, *Shaftesbury and Hutcheson*, English Philosophers, 5 (G.P. Putnam's Sons, 1883)

Fukuyama, Francis, 'The End of History?', *The National Interest*, 16 (1989), pp. 3–18

Furbank, P.N., *Diderot: A Critical Biography* (A.A. Knopf, 1992)

Gallagher, Shaun, *Enactivist Interventions: Rethinking the Mind* (Oxford University Press, 2017)

——, *How the Body Shapes the Mind* (Oxford University Press, 2005)

Gay, John, *Preliminary Dissertation Concerning the Fundamental Principle of Virtue or Morality* (W. Thurlbourn, 1731)

Gendlin, Eugene T., *Focusing* (Random House Publishing Group, 1982)

——, 'Three Assertions About the Body', *Folio*, 12.1 (1993), pp. 21–33

Gérard, Albert, 'Counterfeiting Infinity: "The Eolian Harp" and the Growth of Coleridge's Mind', *Journal of English and Germanic Philology*, 60.3 (1961), pp. 411–422

Gibson, James J., *The Ecological Approach to Visual Perception* (Lawrence Erlbaum Associates, 1986)

Gigante, Denise, *Life: Organic Form and Romanticism* (Yale University Press, 2009)

Giovanni, George di, 'Kant's Critical Legacy: Fichte's Constructivism and Hegel's Discursive Logic', in *The Palgrave Handbook of German Idealism*, ed. by Matthew C. Altman, Palgrave Handbooks (Palgrave Macmillan, 2014), pp. 665–686

Goldstein, Amanda Jo, 'Nerve Poetry and Fiber Art: Biosemiosis and Plasticity in Erasmus Darwin', *Literature Compass*, 17.3–4 (2020), 17:e12552, doi:10.1111/lic3.12552

Goodman, Kevis, 'Reading Motion: Coleridge's "Free Spirit" and Its Medical Background', *European Romantic Review*, 26.3 (2015), pp. 349–356, doi:10.1080/10509585.2015.1028134

Gorodeisky, Keren, '19th Century Romantic Aesthetics', in *The Stanford Encyclopedia of Philosophy*, ed. by Edward N. Zalta and Uri Nodelman, Fall 2016 (Metaphysics Research Lab, Stanford University) <https://plato.stanford.edu/archives/fall2016/entries/aesthetics-19th-romantic/> [accessed 1 June 2021]

Gravil, Richard, 'Tintern Abbey and *The System of Nature*', *Romanticism*, 6.1 (2000), pp. 35–54, doi:10.3366/rom.2000.6.1.35

Greenblatt, Stephen, *Shakespearean Negotiations* (University of California Press, 1988)

Griffiths, Devin, *The Age of Analogy: Science and Literature Between the Darwins* (Johns Hopkins University Press, 2016)

——, 'The Fertile Darwins: Epigenesis, Organicism, and the Problem of Inheritance', *Romanticism on the Net*, no. 66–67 (2016), doi:10.7202/1069959ar

Grob, Alan, *The Philosophic Mind: A Study of Wordsworth's Poetry and Thought 1797–1805* (Ohio State University Press, 1973)

Haeger, J.H., 'Coleridge's Speculations on Race', *Studies in Romanticism*, 13.4 (1974), pp. 333–357

Haraway, Donna J., 'A Cyborg Manifesto: Science, Technology, and Socialist-Feminism in the Late Twentieth Century', in *Manifestly Haraway*, ed. by Cary Wolfe (University of Minnesota Press, 2016), doi:10.5749/minnesota/9780816650477.001.0001

Harding, Anthony John, 'Coleridge as Mentor and the Origins of Masculinist Modernity', *European Romantic Review*, 14.4 (2003), pp. 453–466

——, 'Development and Symbol in the Thought of S.T. Coleridge, J.C. Hare, and John Sterling', *Studies in Romanticism*, 18.1 (1979), pp. 29–48, doi:10.2307/25600166

Harkin, Maureen, 'Introduction', in *The Man of Feeling* (Broadview Press, 2005), pp. 9–38

Harris, William, and Craig Freudenrich, 'How Light Works', *HowStuffWorks*, 1 July 1970 <https://science.howstuffworks.com/light.htm> [accessed 31 May 2019]

Hartford, R.R., 'William King, the Man and His Work', *Theology*, 48.297 (1945), pp. 55–60

Hartley, David, *Observations on Man, His Frame, His Duty, and His Expectations*, 2 vols (S. Richardson, 1749)

——, *Various Conjectures on the Perception, Motion, and Generation of Ideas*, trans. by Robert E.A. Palmer (University of California Press, 1959)

Hartman, Geoffrey H., 'A Poet's Progress: Wordsworth and the "Via Naturaliter Negativa"', *Modern Philology*, 59.3 (1962), pp. 214–224

——, *Wordsworth's Poetry 1787–1814* (Yale University Press, 1971)

Hatch, R.B., 'David Hartley: Freewill and Mystical Associations', *Mosaic*, 7.4 (1974), pp. 29–39

Haven, Richard, 'Coleridge, Hartley, and the Mystics', *Journal of the History of Ideas*, 20.4 (1959), pp. 477–494

Hayden, John, 'Wordsworth, Hartley, and the Revisionists', *Studies in Philology*, 81.1 (1984), pp. 94–118

Hayles, N. Katherine, *How We Became Posthuman: Virtual Bodies in Cybernetics, Literature, and Informatics* (University of Chicago Press, 1999)

——, 'The Materiality of Informatics', *Configurations*, 1.1 (1993), pp. 147–170

Heim, Noel A., and others, 'Hierarchical Complexity and the Size Limits of Life', *Proceedings of the Royal Society B: Biological Sciences*, 284.1857 (2017), doi:10.1098/rspb.2017.1039

Hogarth, William, *The Analysis of Beauty*, ed. by Ronald Paulson (Yale University Press, 1997), Internet Archive <http://archive.org/details/analysisbeauty00hogagoog> [accessed 15 March 2025]

Holbach, Paul Henri Thiry d', *The System of Nature, Or, Laws of the Moral and Physical World*, trans. by Samuel Wilkinson, 3 vols (P. Davison, 1820)

——, *The System of Nature, Or, Laws of the Moral and Physical World*, trans. by Henry D. Robinson, 2 vols (G.W. & A.J. Matsell, 1835)

——, *The System of Nature; or, the Laws of the Moral and Physical World*, trans. by William Hodgson, 4 vols (G. Kearsley, 1797), I, Eighteenth Century Collections Online <link.gale.com/apps/doc/CW0122703088/ECCO?u=asecs&sid=bookmark-ECCO&xid=cf0c90a7&pg=1> [accessed 15 April 2012]

——, *The System of Nature; or, the Laws of the Moral and Physical World*, trans. by William Hodgson, 4 vols (G. Kearsley, 1797), II, Eighteenth Century Collections Online <link.gale.com/apps/doc/CW0122703413/ECCO?u=asecs&sid=bookmark-ECCO&xid=a3ec6a55&pg=1> [accessed 15 April 2012]

Hushagen, Sam, 'Sense Data and the Romantic Myth of the Given', *The Wordsworth Circle*, 53.3 (2022), pp. 311–333, doi:10.1086/720738

Hutcheson, Francis, *An Essay on the Nature and Conduct of the Passions and Affections* (J. Darby and T. Browne, 1728), Eighteenth Century Collections Online <https://link.gale.com/apps/doc/CW0119816864/ECCO?sid=bookmark-ECCO&xid=54b503ee&pg=1> [accessed 22 April 2024]

——, *A System of Moral Philosophy, in Three Books*, 3 vols (A. Millar, 1755)

Hutton, Sarah, 'The Cambridge Platonists', in *The Stanford Encyclopedia of Philosophy*, ed. by Edward N. Zalta and Uri Nodelman, Summer 2024 (Metaphysics Research Lab, Stanford University) <https://plato.stanford.edu/archives/sum2024/entries/cambridge-platonists/> [accessed 21 May 2024]

Ingrassia, Catherine, 'Introduction', in *Anti-Pamela and Shamela* (Broadview Press, 2004)

Ishizuka, Hisaoy, 'Enlightening the Fibre-Woven Body: William Blake and Eighteenth-Century Fibre Medicine', *Literature and Medicine*, 25.1 (2006), pp. 72–92, doi:10.1353/lm.2006.0020

Jackendoff, Ray, *Consciousness and the Computational Mind* (MIT Press, 1987)

Jackson, Noel, 'Rhyme and Reason: Erasmus Darwin's Romanticism', *Modern Language Quarterly*, 70.2 (2009), pp. 171–194

——, *Science and Sensation in Romantic Poetry* (Cambridge University Press, 2008)

Jameson, Fredric, *The Political Unconscious: Narrative as a Socially Symbolic Act* (Cornell University Press, 1983)

Jarvis, Simon, *Wordsworth's Philosophic Song*, Cambridge Studies in Romanticism, 67 (Cambridge University Press, 2007)

Jessa, Tega, 'What Are Photons', *Universe Today*, 19 September 2010 <https://www.universetoday.com/74027/what-are-photons/> [accessed 23 July 2019]

Johnson, Samuel, *A Dictionary of the English Language* (J. and P. Knapton; T. and T. Longman; C. Hitch and L. Hawes; A. Millar; and R. and J. Dodsley, 1755), Internet Archive <http://archive.org/details/johnsons_dictionary_1755> [accessed 30 May 2024]

Kant, Immanuel, *Critique of the Power of Judgment*, ed. by Paul Guyer, trans. by Eric Matthews (Cambridge University Press, 2000)

Kaufman, Robert, 'Legislators of the Post-Everything World: Shelley's "Defence of Adorno"', *ELH*, 63.3 (1996), pp. 707–733

Kim, Jaegwon, *Philosophy of Mind*, ed. by Norman Daniels and Keith Lehrer, Dimensions of Philosophy (Westview Press, 1996)

King, William, *An Essay on the Origin of Evil*, trans. by Edmund Law (W. Thurlbourn, 1731)

King-Hele, Desmond, *Erasmus Darwin* (Charles Scribner's Sons, 1963)

——, *Erasmus Darwin and the Romantic Poets* (Macmillan Press, 1986)

Knappett, Carl, 'The Affordances of Things: A Post-Gibsonian Perspective on the Relationality of Mind and Matter', in *Rethinking Materiality: The Engagement of Mind with the Material World*, ed. by Elizabeth DeMarrais, Chris Gosden, and Colin Renfrew, McDonald Institute Monographs (McDonald Institute for Archaeological Research, 2004), pp. 43–51

Lakoff, George, and Mark Johnson, *Metaphors We Live By* (University of Chicago Press, 1980)

——, *Philosophy in the Flesh: The Embodied Mind and Its Challenge to Western Thought* (Perseus Books Group, 1999)

Lamb, Jonathan, 'Hartley and Wordsworth: Philosophical Language and Figures of the Sublime', *MLN*, 97.5 (1982), pp. 1064–1085

Laughlin, Charles D., John McManus, and Eugene G. D'Aquili, *Brain, Symbol & Experience: Towards a Neurophenomenology of Human Consciousness* (Shambala, 1990)

Lawrence, Christopher, 'Cullen, Brown and the Poverty of Essentialism', *Medical History*, 32.S8 (1988), pp. 1–21, doi:10.1017/S0025727300070770

Lawrence, William, *Lectures on Physiology, Zoology, and the Natural History of Man: Delivered at the Royal College of Surgeons* (Benbow, 1822)

Leadbetter, Gregory, *Coleridge and the Daemonic Imagination* (Palgrave Macmillan, 2011)

LeBuffe, Michael, 'Paul-Henri Thiry (Baron) d'Holbach', in *The Stanford Encyclopedia of Philosophy*, ed. by Edward N. Zalta and Uri Nodelman, Spring 2020 (Metaphysics Research Lab, Stanford University) <https://plato.stanford.edu/archives/spr2020/entries/holbach/> [accessed 3 April 2021]

Leslie, Margaret, 'Mysticism Misunderstood: David Hartley and the Idea of Progress', *Journal of the History of Ideas*, 33.4 (1972), pp. 625–632

Levere, Trevor H., *Poetry Realized in Nature: Samuel Taylor Coleridge and Early Nineteenth-Century Science* (Cambridge University Press, 1981)

Levinas, Emmanuel, *Of God Who Comes to Mind*, trans. by Bettina Bergo (Stanford University Press, 1998)

Levine, Peter A., *What Is Felt Sense?*, dir. by Somatic Experiencing International, YouTube, 2019 <https://www.youtube.com/watch?v=ZBLe84U7AaM> [accessed 28 June 2024]

Levinson, Marjorie, *Thinking Through Poetry: Field Reports on Romantic Lyric* (Oxford University Press, 2018)

——, *Wordsworth's Great Period Poems* (Cambridge University Press, 1986)

Liu, Alan, *Wordsworth: The Sense of History* (Stanford University Press, 1989)

Locke, John, *An Essay Concerning Human Understanding*, 4th edn (Prometheus Books, 1995)

Lokhorst, Gert-Jan, 'Descartes and the Pineal Gland', *The Stanford Encyclopedia of Philosophy*, ed. by Edward N. Zalta and Uri Nodelman, Winter 2021 (Metaphysics Research Lab, Stanford University <https://plato.stanford.edu/archives/win2021/entries/pineal-gland/> [accessed 14 March 2025]

Luhmann, Niklas, *Social Systems*, trans. by John Bednarz, Jr. with Dirk Baecker (Stanford University Press, 1996)

Luhmann, Niklas, *Theories of Distinction: Redescribing the Descriptions of Modernity*, ed. by William Rasch, trans. by Joseph O'Neil, Elliot Schrieber, Kerstin Behnke, and William Whobrey (Stanford University Press, 2002)

Lustila, Getty L., 'John Gay and the Birth of Utilitarianism', *Utilitas*, 30.1 (2018), pp. 86–106, doi:10.1017/S0953820817000115

Magnuson, Paul, '"The Eolian Harp" in Context', *Studies in Romanticism*, 24.1 (1985), pp. 3–20

——, 'The Lake School: Wordsworth and Coleridge', in *The Cambridge Companion to English Literature, 1740–1830*, ed. by Thomas Keymer and Jon Mee, Cambridge Companions to Literature (Cambridge University Press, 2004), pp. 227–243

Makdisi, Saree, *William Blake and the Impossible History of the 1790s* (University of Chicago Press, 2003)

Man, Paul de, *Blindness and Insight: Essays in the Rhetoric of Contemporary Criticism*, 2nd edn (Routledge, 2013)

Marsh, Robert, 'The Second Part of Hartley's System', *Journal of the History of Ideas*, 20.2 (1959), pp. 264–273

Massumi, Brian, *Parables for the Virtual: Movement, Affect, Sensation* (Duke University Press, 2002)

Matlak, Richard, 'Wordsworth's Reading of Zoonomia in Early-Spring', *Wordsworth Circle*, 12.2 (1990), pp. 76–81

McFarland, Thomas, *Coleridge and the Pantheist Tradition* (Clarendon Press, 1969)

McGann, Jerome, *The Poetics of Sensibility: A Revolution in Literary Style* (Oxford University Press, 1996)

——, *The Romantic Ideology: A Critical Investigation* (University of Chicago Press, 1983)

McNaughton, David, 'Butler on Benevolence', in *Joseph Butler's Moral and Religious Thought: Tercentenary Essays*, ed. by Christopher Cunliffe (Clarendon Press, 1992), pp. 269–291

McQueen, Joseph, '"Old Faith Is Often Modern Heresy": Re-Enchanted Orthodoxy in Coleridge's "The Eolian Harp" and "The Rime of the Ancient Mariner"', *Christianity and Literature*, 64.1 (2014), pp. 21–42

Mensch, Jennifer, *Kant's Organicism: Epigenesis and the Development of Critical Philosophy* (University of Chicago Press, 2013)

Merleau-Ponty, Maurice, *Phenomenology of Perception*, trans. by Colin Smith, Routledge Classics (Routledge, 2003)

——, *The Structure of Behavior*, trans. by Alden L. Fisher (Beacon Press, 1963)

Mettrie, Julien Offray de La, *La Mettrie: Machine Man and Other Writings*, trans. and ed. by Ann Thomson, Cambridge Texts in the History of Philosophy (Cambridge University Press, 1996)

Miall, David S., 'The Alps Deferred: Wordsworth at the Simplon Pass', *European Romantic Review*, 9.1 (1998), pp. 87–102, doi:10.1080/10509589808570040

——, 'Wordsworth's "First-Born Affinities": Intimations of Embodied Cognition', *Poetics Today*, 32.4 (2011), pp. 693–715

Mileur, Jean-Pierre, 'Allegory and Irony: "The Rhetoric of Temporality" Re-Examined', *Comparative Literature*, 38.4 (1986), pp. 329–336, doi:10.2307/1770393

Miller, Ashley, '"Striking Passages": Memory and the Romantic Imprint', *Studies in Romanticism*, 50 (2011), pp. 29–53

Milton, John, *Paradise Lost*, ed. by Alastair Fowler, Longman Annotated English Poets, 2nd edn (Longman, 1998)

Mitchell, Robert, *Experimental Life: Vitalism in Romantic Science and Literature* (Johns Hopkins University Press, 2013)

Modiano, Raimonda, *Coleridge and the Concept of Nature* (Florida State University Press, 1985)

Montgomery, Marion, 'Wordsworth and the Ghost in the Mind: The Seductions of Metaphor', *The Georgia Review*, 23.2 (1969), pp. 169–182

Morton, Timothy, 'Of Matter and Meter: Environmental Form in Coleridge's "Effusion 35" and "The Eolian Harp"', *Literature Compass*, 5.2 (2008), pp. 310–335

Muller, Karis, 'Point Atomism, Space and God (1760–80)', *The European Legacy*, 12.3 (2007), pp. 277–292

Newan, Albert, Shaun Gallagher, and Leon De Bruin, eds., *The Oxford Handbook of 4E Cognition*, Oxford Handbooks (Oxford University Press, 2018)

Newton, Isaac, *Opticks: Or, A Treatise of the Reflections, Refractions, Inflections and Colours of Light* (William Innys, 1730)

——, *Sir Isaac Newton's Mathematical Principles of Natural Philosophy and His System of the World*, ed. by Cajori Florian, trans. by Andrew Motte (Cambridge University Press, 1934), Internet Archive <http://archive.org/details/in.ernet.dli.2015.80658> [accessed 19 May 2024]

Oberg, Barbara Bowen, 'David Hartley and the Association of Ideas', *Journal of the History of Ideas*, 37.3 (1976), pp. 441–454, doi:10.2307/2708808

——, 'The Progress Toward the Perfection of Man: David Hartley and the Association of Ideas' (unpublished doctoral dissertation, University of California, 1973)

Oldfield, R.C., and Lady Kathleen Oldfield, 'Hartley's *Observations on Man*', *Annals of Science*, 7.4 (1951), pp. 371–382

O'Malley, Harris, 'Men Really Need to Stop Calling Women Crazy', *Washington Post*, 7 October 2021 <https://www.washingtonpost.com/posteverything/wp/2014/07/09/men-really-need-to-stop-calling-women-crazy/> [accessed 30 June 2024]

Owen, W.J.B., and Jane Worthington Smyser, 'Introduction: General to Postscript, 1835', in *The Prose Works of William Wordsworth*, 3 vols (Oxford University Press, 1974), III, pp. 231–237

Packham, Catherine, *Eighteenth-Century Vitalism: Bodies, Culture, Politics*, Palgrave Studies in the Enlightenment, Romanticism, and Cultures of Print (Palgrave Macmillan, 2012)

Pearsall, Robert Brainard, 'Wordsworth Reworks His Hartley', *The Bulletin of the Rocky Mountain Modern Language Association*, 24.2 (1970), pp. 75–83

Perry, Seamus, *Coleridge and the Uses of Division*, Oxford English Monographs (Oxford: Clarendon Press, 1999)

Pickering, Andrew, *The Mangle of Practice: Time, Agency, and Science* (University of Chicago Press, 2010)

Piper, H.W., *The Active Universe: Pantheism and the Concept of Imagination in the English Romantic Poets* (University of London, Athlone Press, 1962)

Piredda, Giulia, 'The Mark of the Cognitive and the Coupling-Constitution Fallacy: A Defense of the Extended Mind Hypothesis', *Frontiers in Psychology*, 8 (2017), pp. 1664–1078, doi:10.3389/fpsyg.2017.02061

Pittman, Charles, 'An Introduction to a Study of Wordsworth's Reading in Science', *Furman's Studies*, 33.5 (1950), pp. 27–60

Podgórski, Jacek Seweryn, 'From Autopoiesis to Natural Drift Metaphor', *Ecological Questions*, 13 (2014), pp. 81–87

Poetzsch, Markus, 'Between the "Abyss of Idealism" and the Hard Wall of "Reality": Lessons of the Wordsworthian Child', *New Review of Children's Literature and Librarianship*, 20 (2014), pp. 15–25, doi:10.1080/13614541.2014.863642

Porter, Dahlia, *Science, Form, and the Problem of Induction in British Romanticism*, Cambridge Studies in Romanticism, 120 (Cambridge University Press, 2018)

——, 'Scientific Analogy and Literary Taxonomy in Darwin's *Loves of the Plants*', *European Romantic Review*, 18.2 (2007), pp. 213–221, doi:10.1080/10509580701297950

Potkay, Adam, 'Wordsworth's Ethical Thinking', in *The Oxford Handbook of William Wordsworth*, ed. by Richard Gravil and Daniel Robinson, Oxford Handbooks (Oxford University Press, 2015), pp. 679–692

——, *Wordsworth's Ethics* (Johns Hopkins University Press, 2012)

Priestley, Joseph, *Disquisitions Relating to Matter and Spirit* (Kessinger Publishing, 2005)

——, *Hartley's Theory of the Human Mind on the Principle of the Association of Ideas, with Essays Relating to the Subject of It* (J. Johnson, 1775)

——, *The History and Present State of Discoveries Relating to Vision, Light, and Colours* (J. Johnson, 1772)

——, *Institutes of Natural and Revealed Religion*, 2nd edn, 2 vols (Pearson and Rollason for J. Johnson, 1782), I, Eighteenth Century Collections Online <https://link.gale.com/apps/doc/CW0122282972/ECCO?sid=bookmark-ECCO&xid=8d97510e&pg=1> [accessed 23 May 2023]

——, *Institutes of Natural and Revealed Religion*, 2nd edn, 2 vols (Pearson and Rollason for J. Johnson, 1782), II, Eighteenth Century Collections Online <https://link.gale.com/apps/doc/CW0122283468/ECCO?sid=bookmark-ECCO&xid=d254db5f&pg=1> [accessed 23 May 2023]

Priestman, Martin, *The Poetry of Erasmus Darwin: Enlightened Spaces, Romantic Times* (Ashgate, 2013)

Rader, Melvin, 'The Transcendentalism of William Wordsworth', *Modern Philology*, 26.2 (1928), pp. 169–190

Raiger, Michael, 'The Intellectual Breeze, the Corporeality of Thought, and the Eolian Harp', *Coleridge Bulletin: The Journal of the Friends of Coleridge*, 20 (2002), pp. 76–84

Rajan, Tilottama, 'The Unavowable Community of Idealism: Coleridge and the Life Sciences', *European Romantic Review*, 14.4 (2003), pp. 395–416

Ralph Waldo Emerson, 'Nature', in *Nature and Selected Essays* (Penguin, 2003), pp. 35–84

Ramachandran, V.S., 'Mirror Neurons and the Brain in a Vat', *Edge: The Third Culture*, 176 (2006) <Edge Foundation>

Richardson, Alan, *British Romanticism and the Science of the Mind*, Cambridge Studies in Romanticism, 47 (Cambridge University Press, 2001)

——, *The Neural Sublime: Cognitive Theories and Romantic Texts* (Johns Hopkins University Press, 2010)

Ricoeur, Paul, *Freud and Philosophy: An Essay on Interpretation* (Yale University Press, 1970)

Risse, Guenter B., '"Doctor William Cullen, Physician, Edinburgh": A Consultation Practice in the Eighteenth Century', *Bulletin of the History of Medicine*, 48.3 (1974), pp. 338–351

Roberts, Jonathan, 'Wordsworth's Apocalypse', *Literature and Theology*, 20.4 (2006), pp. 361–378

Robertson, Lisa Ann, 'Doctor Beddoes's Democratic Toys: Education, Embodiment, and Empowerment', *Essays in Romanticism*, 26.2 (2019), pp. 115–129

——, '"Swallowed Up In Impression": Humphry Davy's Materialist Theory of Embodied Transcendence and William Wordsworth's "Tintern Abbey"', *European Romantic Review*, 26.5 (2015), pp. 591–614, doi:10.1080/10509585.2015.1070343

——, '"Time + Space": Coleridge's Phenomenological Organicism', *Coleridge Bulletin*, n. s., 62 (2024), pp. 3–19

Roca, Julius, 'William Cullen (1710–1790) and Robert Whytt (1714–1766) on the Nervous System', in *Brain, Mind and Medicine: Essays in Eighteenth-Century Neuroscience*, ed. by Harry Whitaker, C.U.M. Smith, and Stanley Finger (Springer, 2007), pp. 85–98

Rosen, Michael, 'Freedom in German Idealism', *Harvard University*, May 2007 <https://scholar.harvard.edu/michaelrosen/publications/freedom-germanidealism> [accessed 11 March 2025]

Ross, Catherine E., 'Coleridge's "The Eolian Harp"', *The Explicator*, 91.1 (2002), pp. 17–20

Rowland, Ann Wierda, *Romanticism and Childhood: The Infantilization of British Literary Culture*, Cambridge Studies in Romanticism, 93 (Cambridge University Press, 2015)

Rowlands, Mark, *The New Science of the Mind: From Extended Mind to Embodied Phenomenology* (Bradford Books, 2010)

Ruderman, D.B., *The Idea of Infancy in Nineteenth-Century British Poetry: Romanticism, Subjectivity, Form* (Routledge, 2016)

Ruston, Sharon, *Creating Romanticism: Case Studies in the Literature, Science and Medicine of the 1790s*, Palgrave Studies in the Enlightenment, Romanticism, and Cultures of Print (Palgrave Macmillan, 2013)

——, '"Natural Enemies in Science, as Well as in Politics": Romanticism and Scientific Conflict', *Romanticism*, 11.1 (2005), pp. 70–83

Ryle, Gilbert, *The Concept of Mind* (1949)

Rynasiewicz, Robert, 'Newton's Views on Space, Time, and Motion', *The Stanford Encyclopedia of Philosophy*, ed. by Edward N. Zalta and Uri Nodelman, Summer 2014 (Metaphysics Research Lab, Stanford University) <http://plato.stanford.edu/archives/sum2014/entries/newton-stm/> [accessed 25 November 2016]

Rzepka, Charles, 'From Relics to Remains: Wordsworth's "The Thorn" and the Emergence of Secular History', *Romanticism on the Net*, no. 31 (2003), doi:10.7202/008696ar

Sanders, C. Richard, 'The Background of Carlyle's Portrait in *The Life of John Sterling*', *Bulletin of the John Rylands Library*, 55.2 (1973), pp. 434–458

Sangiacomo, Andrea, and Desmond Clarke, 'Louis de La Forge', in *The Stanford Encyclopedia of Philosophy*, ed. by Edward N. Zalta and Uri Nodelman, Summer 2020 (Metaphysics Research Lab, Stanford University) <https://plato.stanford.edu/archives/sum2020/entries/la-forge/> [accessed 1 June 2022]

Sarafianos, Aris, 'The Contractility of Burke's Sublime and Heterodoxies in Medicine and Art', *Journal of the History of Ideas*, 69.1 (2008), pp. 23–48, doi:10.1353/jhi.2008.0005

——, 'Pain, Labor, and the Sublime: Medical Gymnastics and Burke's Aesthetics', *Representations*, 91.1 (2005), pp. 58–83, doi:10.1525/rep.2005.91.1.58

Savarese, John, *Romanticism's Other Minds: Poetry, Cognition, and the Science of Sociability* (Ohio State University Press, 2020)

Savarese, Ralph James, 'Cognition', in *Keywords for Disability Studies*, ed. by Rachel Adams, Benjamin Reiss, and David Serlin (New York University Press, 2015), pp. 40–42

Scarry, Elaine, *Dreaming by the Book* (Princeton University Press, 2001)

Scheuerle, William H., 'A Reexamination of Coleridge's "The Eolian Harp"', *SEL: Studies in English Literature, 1500–1900*, 15.4 (1975), pp. 591–599

Schmid, Thomas H., '"Diminished Impressibility": Addiction, Neuroadaptation, and Pleasure in Coleridge', in *Romanticism and Pleasure*, ed. by Thomas H. Schmid and Michelle Faubert (Palgrave Macmillan, 2010), pp. 165–186

Schofield, Robert, 'Joseph Priestley on Sensation and Perception', in *Studies in Perception: Interrelations in the History of Philosophy and Science*, ed. by Peter K. Machamer and Robert G. Turnbull (Ohio State University Press, 1978), pp. 336–354

Sedgwick, Peter H., *The Development of Anglican Moral Theology, 1680–1950* (Brill, 2024)

Seward, Thomas, 'Untitled', *The Gentleman's Magazine*, 54.2 (1784), p. 87

Sha, Richard C., 'Blakean Experience and the Hard Problem of Consciousness Revisited', in *Romanticism and Consciousness, Revisited*, ed. by Richard C. Sha and Joel Faflak (Edinburgh University Press, 2022), pp. 94–117

——, *Imagination and Science in Romanticism* (Johns Hopkins University Press, 2018)

——, 'Romantic Physiology and the Work of Romantic Imagination: Hypothesis and Speculation in Science and Coleridge', *European Romantic Review*, 24.4 (2013), pp. 403–419

Sha, Richard C., and Joel Faflak, 'Introduction: *Romanticism and Consciousness* Redux', in *Romanticism and Consciousness, Revisited*, ed. by Richard C. Sha and Joel Faflak (Edinburgh University Press, 2022), pp. 1–14

Shaftesbury, Anthony Ashley Cooper, *Characteristics of Men, Manners, Opinions, Times, Etc*, ed. by John Mackinnon Robertson, 2 vols (G. Richards, 1900), II, Internet Archive <http://archive.org/details/characteristicso02shaf> [accessed 27 June 2024]

——, *Several Letters Written by a Noble Lord to a Young Man at the University* (J. Roberts, 1716)

Shaver, Chester L., and Alice C. Shaver, *Wordsworth's Library: A Catalogue* (Garland Publishing, 1979)

Shaw, Devin Zane, 'The "Keystone" of the System: Schelling's Philosophy of Art', in *The Palgrave Handbook of German Idealism*, ed. by Matthew C. Altman, Palgrave Handbooks (Palgrave Macmillan, 2014), pp. 518–538

Shelley, Percy Bysshe, 'Mont Blanc', in Mary Shelley and Percy Bysshe Shelley, *History of a Six Weeks' Tour Through a Part of France, Switzerland, Germany and Holland: With Letters Descriptive of a Sail Round the Lake of Geneva, and of the Glaciers of Chamouni* (T. Hookham, and C. and J. Ollier, 1817), pp. 171–183

Smith, Adam, *The Theory of Moral Sentiments*, 4th edn (W. Strahan, J. and F. Rivingtong, W. Johnson, T. Longman, and T. Cadell, 1774), Eighteenth Century Collections Online <link.gale.com/apps/doc/CW0122297518/ECCO?u=asecs&sid=bookmark-ECCO&xid=f2003db8&pg=1> [accessed 2 May 2008]

Smith, C.U.M., 'Coleridge's "Theory of Life"', *Journal of the History of Biology*, 32.1 (1999), pp. 31–50

——, 'David Hartley's Newtonian Neuropsychology', *History of Behavioral Sciences*, 23.2 (1987), pp. 123–136

——, 'Julien Offray de La Mettrie (1709–1751)', *Journal of the History of the Neurosciences*, 11.2 (2002), pp. 110–124, doi:10.1076/jhin.11.2.110.15188

Snyder, Alice Dorothea, *Coleridge on Logic and Learning* (New Haven, 1929), HathiTrust Digital Library <http://hdl.handle.net/2027/mdp.39015074764096> [accessed 1 June 2023]

Stanback, Emily B., *The Wordsworth-Coleridge Circle and the Aesthetics of Disability*, Palgrave Studies in Literature, Science and Medicine (Palgrave Macmillan, 2016)

Steigerwald, Joan, *Experimenting at the Boundaries of Life: Organic Vitality in Germany around 1800* (University of Pittsburgh Press, 2019)

Sutcliffe, F.E., 'Introduction', in René Descartes, *Discourse on Method and the Meditations* (Penguin Books Ltd, 1971)

Tennant, Bob, *Conscience, Consciousness, and Ethics in Joseph Butler's Philosophy and Ministry* (Boydell & Brewer, 2011)

Tennyson, Alfred, *In Memoriam*, ed. by Erik Gray, 2nd edn (W.W. Norton, 2004), Internet Archive <http://archive.org/details/inmemoriamauthor0000tenn> [accessed 19 January 2024]

Thomas, Keith G., *Wordsworth and Philosophy: Empiricism and Transcendentalism in the Poetry* (UMI Research Press, 1989)

Thompson, Evan, *Mind in Life: Biology, Phenomenology, and the Science of Mind* (Harvard University Press, 2007)

Todd, Janet, *Sensibility: An Introduction* (Methuen, 1986)

Trower, Shelley, 'Nerves, Vibration and the Aeolian Harp', *Romanticism and Victorianism on the Net*, 54 (2009)

Tsouyopoulos, Nelly, 'The Influence of John Brown's Ideas in Germany', *Medical History*, 32.S8 (1988), pp. 63–74, doi:10.1017/S0025727300070800

Vallins, David., *Coleridge and the Psychology of Romanticism: Feeling and Thought* (Macmillan, 2000)

Varela, Francisco J., 'Autopoiesis and a Biology of Intentionality', in *Autopoiesis and Perception: Proceedings of a Workshop Held at Dublin City University on August 25th & 26th*, ed. by Finbarr (Barry) V. McMullin and Noel Murphy (Dublin City University, School of Electronic Engineering Technical Report, 1992), pp. 4–14

Varela, Francisco J., Evan Thompson, and Eleanor Rosch, *The Embodied Mind: Cognitive Science and Human Experience* (MIT Press, 1991)

Vickers, Neil, *Coleridge and the Doctors: 1795–1806* (Oxford University Press, 2004)

———, 'Coleridge, Thomas Beddoes and Brunonian Medicine', *European Romantic Review*, 8.1 (1997), pp. 47–94

Vrijders, Dries, 'History, Poetry, and the Footnote: Cleanth Brooks and Kenneth Burke on Keats's "Ode on a Grecian Urn"', *New Literary History*, 42.3 (2011), pp. 537–552

Walker, John, *Lectures on Geology: Including Hydrography, Mineralogy, and Meteorology, with an Introduction to Biology*, ed. by Harold W. Scott (University of Chicago Press, 1966)

Walker, Peter, 'Cross-Sensory Correspondences: A Theoretical Framework and Their Relevance to Music', *Psychomusicology*, 26.2 (2016), pp. 103–116

Walsh, Philip J., 'The Concept of Motivation in Merleau-Ponty: Husserlian Sources, Intentionality, and Institution', *Journal of the History of Philosophy*, 60.2 (2022), pp. 303–336

Ward, Dave, 'Enactivism', *A History of Distributed Cognition*, November 2024 (University of Edinburgh) <http://www.hdc.ed.ac.uk/seminars/enactivism> [accessed 7 February 2015]

Warner, Rebecca, *Original Letters, from Richard Baxter, Matthew Prior, Lord Bolingbroke, Alexander Pope, Dr. Cheyne, Dr. Hartley, Dr. Samuel Johnson, Mrs. Montague, Rev. William Gilpin, Rev. John Newton, George Lord Lyttleton, Rev. Claudius Buchanan, Etc. Etc.* (Richard Cruttwell, 1817) <https://www.google.com/books/edition/Original_Letters_from_Richard_Baxter_Mal/-RKZT6wUA98C?hl=en&gbpv=0> [accessed 24 May 2010]

Webster, Suzanne, *Body and Soul in Coleridge's Notebooks, 1827–1834: 'What Is Life?'* (Palgrave Macmillan, 2010)

Wedgwood, Thomas, Victoria & Albert Museum, Wedgwood Collection Archives

Wehrs, Donald R., 'Epilogue: Literary Theory and Cognitive Studies', in *Cognition, Literature, and History*, ed. by Mark J. Bruhn and Donald R. Wehrs (Routledge, 2014), pp. 243–252

Wellman, Kathleen, 'Medicine as a Key to Defining Enlightenment Issues: The Case of Julien Offray de La Mettrie', *Studies in Eighteenth-Century Culture*, 17 (1987), pp. 75–89

Wendling, Ronald C., 'Coleridge and the Consistency of "The Eolian Harp"', *Studies in Romanticism*, 8.1 (1968), pp. 26–42

'Wildered, Adj.', in *Oxford English Dictionary Online*, n.d. <http://www.oed.com.ezproxy.usd.edu/view/Entry/228999> [accessed 13 December 2016]

Williams, Raymond, *Marxism and Literature* (Oxford University Press, 1977)

———, *Problems in Materialism and Culture: Selected Essays* (Verso, 1980)

Willmott, Glenn, 'Blood in the Blue Lagoon: Byron and the Politics of Wonder', in *Awake to Love and Beauty: Proceedings from a Conference in Honour of George Whally*, ed. by Michael John DiSanto, and others (Algoma University, 2016), pp. 65–75

Wolfe, Cary, 'Meaning as Event-Machine, or Systems Theory and "The Reconstruction of Deconstruction": Derrida and Luhmann', in *Emergence and Embodiment: New Essays on Second-Order Systems Theory*, ed. by Bruce Clarke and Mark B. N. Hansen, Science and Cultural Theory (Duke University Press, 2009), pp. 220–245

——, *What Is Posthumanism?*, Posthumanities, 8 (University of Minnesota Press, 2010)

Wolfe, Charles T., 'Determinism/Spinozism in the Radical Enlightenment: The Cases of Anthony Collins and Denis Diderot', in *Boundaries in the Eighteenth Century*, ed. by Pasi Ihalainen, and others (Finnish Society for Eighteenth-Century Studies, 2007), pp. 37–51

Wordsworth, William, *The Correspondence of Henry Crabb Robinson with the Wordsworth Circle, 1808–1866*, ed. by Edith J. Morley, 2 vols (Clarendon Press, 1927)

——, 'Essay, Supplementary to the Preface', *The Prose Works of William Wordsworth*, ed. by W.J.B. Owen and Jane Worthington Smyser, 3 vols (Clarendon Press, 1974), III, pp. 62–84

——, *The Excursion*, ed. by Sally Bushell, James A. Butler, and Michael C. Jaye, The Cornell Wordsworth (Cornell University Press, 2007)

——, 'A Letter to the Bishop of Llandaff', *The Prose Works of William Wordsworth*, ed. by W.J.B. Owen and Jane Worthington Smyser, 3 vols (Clarendon Press, 1974), I, pp. 29–49

——, 'Lines Written a Few Miles above Tintern Abbey, on Revisiting the Banks of the Wye During a Tour, July 13, 1798', in *Lyrical Ballads, and Other Poems, 1797–1800 by William Wordsworth*, ed. by James Butler and Karen Green, The Cornell Wordsworth (Cornell University Press, 1992), pp. 116–120

——, 'Ode', in *Poems, in Two Volumes, and Other Poems 1800–1807*, ed. by Jared Curtis, The Cornell Wordsworth (Cornell University Press, 1983), pp. 269–277

——, *The Poetical Works of William Wordsworth*, ed. by Ernest de Selincourt, 5 vols (Clarendon Press, 1963)

——, 'Postscript, 1835', *The Prose Works of William Wordsworth*, ed. by W.J.B. Owen and Jane Worthington Smyser, 3 vols (Clarendon Press, 1974), III, pp. 240–274

——, 'Preface', *The Prose Works of William Wordsworth*, ed. by W. J. B. Owen and Jane Worthington Smyser, 3 vols (Clarendon Press, 1974), I, pp. 118–159

——, 'Preface of 1815', *The Prose Works of William Wordsworth*, ed. by W. J. B. Owen and Jane Worthington Smyser, 3 vols (Clarendon Press, 1974), III, pp. 25–39

——, *The Prelude: 1799, 1805, 1850*, ed. by Jonathan Wordsworth, M.H. Abrams, and Stephen Gill (W.W. Norton, 1979)

Wordsworth, Jonathan, *The Music of Humanity: A Critical Study of Wordsworth's 'Ruined Cottage' Incorporating Texts from a Manuscript of 1799–1800* (Harper & Row Print, 1969)

——, 'Wordsworth's "Borderers"', in *English Romantic Poets: Modern Essays in Criticism*, 2nd edn, ed. by M.H. Abrams (Oxford University Press, 1975), pp. 170–187

Wu, Duncan, *Wordsworth's Reading 1770–1799* (Cambridge University Press, 1993)

Yolton, John W., *Thinking Matter: Materialism in Eighteenth-Century Britain* (University of Minnesota Press, 1983)

Yousef, Nancy, 'After-Affects and Second Thoughts: Wordsworth, Eliot, and the Forms of Emotional Thinking', in *Romanticism and Consciousness, Revisited*, ed. by Richard C. Sha and Joel Faflak, Edinburgh Critical Studies in Romanticism (Edinburgh University Press, 2022), pp. 144–166

Index

Printed and bound by CPI Group (UK) Ltd, Croydon, CR0 4YY
13/09/2026
14961098-0005